'Scott Carney is so curious about getting to the truth of things that he is willing to endure great pain and suffering to get there. While investigating the controversial methods of Wim Hof and others operating on the scientific fringe, Carney entered a skeptic yet emerged a true believer. In What Doesn't Kill Us, readers get to follow him along on his transformational journey, and the insights are truly fascinating. Informative, fun, and with a healthy degree of danger, this is a book for the adventurer in all of us.'
Gabriel Reece , Co-founder of XPT (Extreme Performance Training)

'The further we get from the harsh environmental conditions that once threatened our existence, the more we need them. I see this every weekend at a Spartan Race somewhere in the world. Millions of otherwise sane people line up to suffer and push themselves to their physical limits, and it feels good. *What Doesn't Kill Us* is a fascinating investigation into the innate urge that drives these people and reveals how some have managed to use environmental conditioning to accomplish truly extraordinary things.'
Joe De Sena, founder, Spartan Race

'As Navy SEALS, we live by the mantra, "what doesn't kill us only makes us stronger". We would hear this phrase and repeat it, but we never had any proof that it was factual. Yet through comprehensive study, Scott Carney has brilliantly documented how engaging in environmental conditioning, breathing, meditation, and other techniques can actually make us physically and mentally stronger. *What Doesn't Kill Us* is a fascinating book that will captivate all who read it, and that will be of immense value to those in the military, those who are active in sports, and those who seek an alternate means of developing greater mental and physical strength.'
Don D. Mann, author of *Inside SEAL Team SIX*

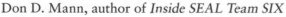

ALSO BY SCOTT CARNEY:

THE RED MARKET

A DEATH ON DIAMOND MOUNTAIN

Vic,
cheers for all of
your help. It has been
extremely appreciated!
see you around
Rob

WHAT DOESN'T KILL US

HOW FREEZING WATER, EXTREME ALTITUDE, AND ENVIRONMENTAL CONDITIONING WILL RENEW OUR LOST EVOLUTIONARY STRENGTH

SCOTT CARNEY

FOREWORD BY WIM HOF

SCRIBE

Melbourne • London

Scribe Publications
18–20 Edward St, Brunswick, Victoria 3056, Australia
2 John Street, Clerkenwell, London, WC1N 2ES, United Kingdom

Published in Australia and the United Kingdom by Scribe 2017

The information in this book is meant to supplement, not replace, proper exercise training. All
forms of exercise pose some inherent risks. The editors and publisher advise readers to take full
responsibility for their safety and know their limits. Before practising the exercises in this book,
be sure that your equipment is well maintained, and do not take risks beyond your level of experience, aptitude, training, and fitness. The exercise and dietary programs in this book are not
intended as a substitute for any exercise routine or dietary regimen that may have been prescribed by your doctor. As with all exercise and dietary programs, you should get your doctor's
approval before beginning.

Mention of specific companies, organisations, or authorities in this book
does not imply endorsement by the author or publisher, nor does mention of
specific companies, organisations, or authorities imply that they endorse
this book, its author, or the publisher.

Internet addresses and telephone numbers given in this book
were accurate at the time it went to press.

Book design by Joanna Williams

Printed and bound in the UK by CPI Group (UK) Ltd, Croydon CR0 4YY

Scribe Publications is committed to the sustainable use of natural resources
and the use of paper products made responsibly from those resources.

9781925321999 (ANZ edition)
9781911344193 (UK edition)
9781925548082 (e-book)

CiP records for this title are available from the British Library.

scribepublications.com.au
scribepublications.co.uk

FOR LAURA KRANTZ

⚠ WARNING

This book is intended to be a journalistic investigation into the limits and possibilities of the human body. No one should attempt any of these methods or practices without appropriate experience, training, fitness level, doctor approval, and supervision—and even then, readers must be aware that these practices are inherently dangerous and could result in grave harm or death.

CONTENTS

Instead of softening their feet with shoe or sandal, his rule was to make them hardy through going barefoot. This habit, if practiced, would, as he believed, enable them to scale heights more easily and clamber down precipices with less danger. In fact, with his feet so trained the young Spartan would leap and spring and run faster unshod than another shod in the ordinary way. Instead of making them effeminate with a variety of clothes, his rule was to habituate them to a single garment the whole year through, thinking that so they would be better prepared to withstand the variations of heat and cold.

—*Xenophon of Sparta, 431–354 BC*

The daily cold plunge does not necessarily place a man "next to the Gods," as he so frequently thinks it does. Such cold-plungers are often very proud of their accomplishment and sneer at those who do not take this daily treatment, and the plunger is likely to "thank God that he is not like other men." Very many times daily cold plunges or cold showers are harmful, especially to those who are underweight or are losing too much weight.

—*Journal of the American Medical Association, 1914*

FOREWORD

NATURE GAVE US the ability to heal ourselves. Conscious breathing and environmental conditioning are two tools that everyone can use to control their immune system, better their moods, and increase their energy. I believe that anyone can tap into these unconscious processes and eventually control their autonomic nervous system. It's a grand claim and some people rightly read my conviction and enthusiasm with incredulity. Skepticism is good and allows the truth to come out. But I wasn't quite sure I was ready for Scott Carney, because he was the most skeptical of all. He came to Poland to prove to the world that I was a fake.

I run a small training center in the cold Karkonosze Mountains, where I teach people to use the snow and ice to break into their deepest physiology. Most everyone comes here motivated to learn. But Scott was different. He is an anthropologist and investigative journalist who is in the habit of asking questions until he gets to the truth of things. The moment I met him at the airport terminal, I knew that it was not going to be an easy week.

I first got to know his analytical mind over a game of chess. We stayed up late probing each other's defenses and staking our claims on the board while also talking about what it means to learn to love the cold. He won the game. But he also made a pact with me to give the training a fair shot.

The next day, he began to learn the techniques for himself. This is a man who had just come from Los Angeles surf country, where the weather is always warm. But he would learn the breathing and lie almost naked in the snow with the rest of the group. I don't imagine that it was something he ever actually *wanted* to do. Yet two days after we met he stood barefoot in the snow, no doubt feeling the primordial powers within.

The Western-lifestyle makes it all-too-easy to take nature for granted. All mammals share the same underlying physiology, but somehow humans are so caught up thinking big thoughts with their big minds that they've come to believe that they're different from everything else around them. Sure we can build skyscrapers, fly airplanes and simply turn up the thermostat to combat the cold, but it turns out that the technologies that we believe are our greatest strengths are also our most tenacious crutches. The things we have made to keep us comfortable are making us weak.

But it only takes a handful of days to begin kicking the dependence on comfort. Conscious breathing and mental focus can jump-start a chemical change to alkalize the body, while immersion in cold water creates a mental and physical mirror for seeing ourselves in a state of fight-or-flight. Feeling that change is powerful.

For the next few years Scott and I kept in touch through e-mail as I discovered new ways to make the method accessible to anyone. He got six pages in the July 2014 issue of *Playboy* magazine! And there I was, an almost naked man in the pages of *Playboy,* spreading the message that breathing exercises can activate the brain stem, where freeze, fight, flight and fuck are the body's most basic instincts. Shortly after that, new studies began to appear in scientific journals containing proof that the techniques worked. Scott knew that it was time to write a book. It would be a simple and effective examination. No speculation. He just did it.

He stayed for three weeks with me in my place in the Netherlands. I believe he discovered that I'm not dogmatic, but simply determined in my goal that everybody is able to become more human.

Earlier this year he set his mind to climbing Mount Kilimanjaro with me. And I hope I'm not giving away any spoilers, but we bloody did it in record time—just 28 hours to the summit. There are no stories or fibs, just real testimonials of what people can achieve if they put their bodies and minds to it.

It's time to bring Mother Nature's power back into our awareness. We are warriors seeking strength and happiness for everyone. Together we regain what we've lost. In other words, there's nothing else to say other than "Breathe, motherfucker."

Love,
Wim Hof
Stroe, Netherlands, April 28, 2016

PREFACE

BURNING UP

OUR LINE OF headlamps cuts through the inky black African night, illuminating patches of a loose rubble path. Aluminum poles and hiking boots crunch the dirt as the group moves ever northward toward a volcanic chunk of rock that claims the lives of about eight mountaineers a year. Our breathing is harsh and rhythmic, as if we are trapped in a room while the air is being sucked away. It sounds as if any given lungful might be our last. We trudge forward in focused unison until the fingers of an orange dawn light grab at the horizon to pull the night away. Now the outline of a mountain peak begins to define itself. At first it is only a dark purple absence of stars in a pinprick sky, but as the heavens shuffle off night's embrace, sunlight sets the glacier ablaze like a beacon.

Kilimanjaro.

The tallest mountain in Africa rises up out of the sun-drenched savannah to a place high above the clouds. There, winds topping 100 miles an hour scour what is likely the only indigenous ice on the continent. It's the first time we've seen it this close, and I can't decide whether I'm excited or terrified. For the past 20 hours the peak hid behind clouds and the mountain's own towering foothills, but now the massive slab of igneous rock is no longer an idea to conjure up in our minds, but a deadly, real-life obstacle. Our gradual 15.5-mile ascent from the park gate will come to an abrupt halt in a few miles, where the base of the volcanic cone rockets upward out of the basin and into a barren and inhospitable wasteland. Devoid of life, and home to only a moonlike base camp, that point will be the start of the greatest challenge of my life—one that will push me to the very limit of human endurance. While thousands of tourists attempt to summit the mountain every year, they tend to do it in easy stages and while wearing the most advanced mountaineering equipment. We will reach the top of the mountain at a record-bending pace with no acclimation to the altitude, on almost no food, little sleep, and, most strikingly, no cold weather gear. I'm wearing only boots, a bathing suit, a wool cap, and a backpack containing some emergency gear and water. My chest is bare to the frigid air.

One of the guides watches me warily from beneath his full thermal getup until finally he can't hold his silence anymore. "Please put something on," he says, concerned by the display of skin.

It's a sensible request. Even with the sun coming up, the temperature is well below freezing, and it is only going to get colder as we ascend higher.

What he doesn't know is that the cold is the least of my concerns. In fact it's kind of the point. My skin feels like armor that the temperature cannot penetrate. Partly it's because I'm working so hard to go uphill that my body has more heat than it knows what to do with, but on another level—one that I'm still trying to wrap my mind around—it's because I just won't let it in. Either way, I'm sweating, not shivering. But there's another challenge that poses a much more pernicious problem, one that could put the whole expedition out of commission.

Reasonable people take 5 to 10 days to reach the top of Kilimanjaro, climbing in slow and deliberate stages along the route so that their bodies can generate enough new red blood cells to compensate for the decrease in oxygen as the altitude increases. But we are not reasonable people. Our rather audacious plan is to make the summit in 2 days. At that pace there is no time for acclimatization. At just above 13,000 feet—and only about two-thirds of the way up—the air is already thin enough to send some unacclimatized people into a spiraling cascade of headaches, convulsions, and sometimes even death. The condition has already created two empty spots in our procession. One belonged to a nearly 7-foot tall Dutchman who spent 10 minutes vomiting up his breakfast this morning and then couldn't stop stumbling with every step. And then there was the owner of a string of Holland's famous marijuana vending "coffee shops," who had so little oxygen in her blood last night that her limbs simply ceased to function.

Mountain sickness can cut down even the most robust athletes. The military has been so perplexed by the problem that when they send special forces units into high altitude combat zones—the type that are all too common in Afghanistan—they have to account for a predictable percentage of their soldiers being incapacitated by the lack of oxygen. So far, the only solution has been to send in additional fighters on every mission. If we are going strictly by the numbers, the prognosis for our group is grim. A day before our departure, a senior scientist at an Army research center that focuses on environmental risks calculated that three-quarters of us would fall to a similar fate as the two we have already lost. The Army isn't alone in being certain that most of us are going to fail. Just before I left, a journalist who spends much of his time topping Colorado's 14,000-foot peaks confided to my wife that he was pretty sure that I would never make it to the top.

What is hard to communicate to the rest of the world is that what we are doing on the mountain is not a stunt or a suicide mission. The lack of clothing, the altitude, and the pace are actually part of an experiment to understand one of the most pressing questions in the modern world today: Has dependence on technology made us weak? Just about every person I know, from the skeptical journalist in Colorado to the US Army scientist to the guide by my side, envelops themselves in a cocoon of technology that keeps them safe, warm, and helps them endure the natural variations of our planet. In the past six million years of human evolution our ancestors mounted expeditions across frozen mountains and parched deserts with only a whisper of technology to aid them. While they might not have aimed to get up this particular mountain, they surely crossed the Alps and the Himalayas, navigated oceans, and populated the New World. What power did they have that we have lost? More important, is it possible to get it back? The underlying hypothesis of this expedition is that when humans outsource comfort and endurance they inadvertently make their bodies weaker, and that simply reintroducing some common environmental stresses to their daily routines can bring back some of that evolutionary vigor. Every person in this line of wobbling headlamps is potentially putting their life on the line to test the theory. We also know that along with the act of conditioning ourselves over time, there's also a simple mind-set and mental fortitude that seems to unlock a biological power to heat our bodies.

I suck in a cool breath of air and focus my eyes on the blazing orange rock in front of me. I exhale a low guttural roar, like a dragon just waking from a thousand-year slumber. I feel the energy begin to build. The rhythm of the air quickens. My toes start to tingle inside my hiking boots. The world starts to brighten in my vision as if there are two dawns working at the same time—one tied to the rising of the sun, the other in the depths of my own mind. A coil of heat starts behind my ears like someone has lit a fuse. It arcs across my shoulders and down the curve of my spine. There's no point in checking the temperature. It's well below freezing and I'm already burning up.

INTRODUCTION

AN ODE TO A JELLYFISH

I DON'T LIKE to suffer. Nor do I particularly want to be cold, wet, or hungry. If I had a spirit animal it would probably be a jellyfish floating in an ocean of perpetual comfort. Every now and then I'd snack on some passing phytoplankton, or whatever it is that jellyfish snack on, and I'd use the tidal forces of the ocean to keep me at the optimal depth. If I were lucky enough to have come into the world as a *Turritopsis dohrnii,* the so-called "immortal jellyfish," then I wouldn't even have to worry about death. When my last days approached, I could simply shrivel into a ball of goo and reemerge a few hours later as a freshly minted juvenile version of myself. Yes, it would be awesome to be a jellyfish.

Unfortunately, as it turns out, I am not an amorphous blob of sea-goop. As a human I am merely the most recent iteration of several hundred million years of evolutionary development from the time we were all just muck in a primordial soup. Most of those previous generations had it pretty rough. There were predators to outwit, famines to endure, species-ending cataclysms to evade, and an ever-changing struggle to survive in outright hostile environments. And, let's be real, most of those would-be ancestors died along the way without passing on their genes.

Evolution is a continual battle waged through generations of minute mutations where only particularly fit or lucky creatures outperform hapless genetic dead ends. The body we have today hasn't stopped evolving, but I still think if we peel back all the eons of changes that brought us here today that we will still find a little bit of jellyfish at the very core of our beings.

This is because we have a nervous system that is almost perfectly attenuated for homeostasis: the effortless state where the environment meets every physical need. Our nervous system automatically responds to challenges in the world around us—triggering muscle contractions, releasing hormones, modulating body temperature, and performing a million other tasks that give us an edge in a particular moment.

But barring an urgent need for survival the human body is perfectly content to simply rest and do nothing. Doing things, *doing anything,* requires a certain amount of energy, and our bodies would rather save up that energy just in case they need it later. The great bulk of these bodily functions lie just beneath our conscious thoughts, but if whatever motivates our nervous system *could* express itself, it would probably maintain that the body that it is responsible for would best tick by admirably well in a state of perpetual and stressless comfort.

But what is comfort? It's not really a feeling as much as it is an absence of things that aren't comfortable. Our species might never have survived necessary but arduous treks across scorching deserts or over frigid mountain peaks if there weren't the promise of some physical reward at the end of the journey. We sate our thirst, don layers of clothing on cold winter days, and clean our bodies because that yearning for comfort is hardwired into our brains. It's what Freud called the "pleasure principle."

The programming that makes us gluttons for the easy life didn't emerge out of nowhere. Aside from my jellyfish spirit animal, almost every organism struggles against the environment that it inhabits. Every biological adaptation that makes life incrementally easier came through the glacial progress of natural selection, when two animals were able to pass favorable traits onto their descendants. Yet evolution requires more than a biological duty that culminates in a moment of intense passion; it needs the cumulative luck, motivations, and skill of individual creatures to use their biological abilities to the fullest. Every creature, whether it is an amoeba or a great ape, needs motivation to overcome the challenges of the world around it. Comfort and pleasure are the two most powerful and immediate rewards that exist.

Anatomically modern humans have lived on the planet for almost 200,000 years. That means your officemate who sits on a rolling chair beneath fluorescent lights all day has pretty much the same basic body as the prehistoric caveman who made spear points out of flint to hunt antelope. To get from there to here humans faced countless challenges as we fled predators, froze in snowstorms, sought shelter from the rain, hunted and gathered our food, and continued breathing despite suffocating heat. Until very recently there was never a time when comfort could be taken for granted—there was always a balance between the effort we expended and the downtime we earned. For the bulk of that time we managed these feats without even a shred of what anyone today would consider modern technology. Instead, we had to be strong to survive. If your pasty-skinned officemate had the ability to travel back in time and meet one of his prehistoric ancestors it would be a very bad idea for him to challenge that caveman to a footrace or a wrestling match.

Over the course of hundreds of thousands of years humans invented some things that made life easier—fire, cooking, stone tools, fur skins, and foot bindings—but we were still largely at the mercy of nature. About 5,000 years ago, at the dawn of recorded history, things got a little easier still as we domesticated various animal species to do work for us, built better shelters, and carried more sophisticated gear. As human culture advanced at least it all was getting incrementally easier. Even so, being a human was not exactly carefree. Each age let us depend more on our ingenuity and less on our basic biology until technological progress was poised to outpace evolution itself. And then, sometime in the early 1900s, our technological prowess became so powerful that it broke our fundamental biological links to the world around us. Indoor plumbing, heating systems, grocery stores, cars, and electric lighting now let us control and fine-tune our environment so thoroughly that many of us can live in what amounts to a perpetual state of homeostasis.

It doesn't matter what the weather is like outside—scorching heat, blizzards, thunderstorms, or just fine summer days—a person can wake up long past when the sun rises, eat a breakfast chock-full of fruits flown in from a climate halfway across the globe, head to work in a temperature-controlled car, spend the day in an office, and come home without ever feeling the outside air for more than a few minutes. Modern humans are the very first species since the jellyfish that can almost completely ignore their natural obstacles to survival.

Yet comfort's golden age has a hidden dark side. While we can imagine what a difficult environment might feel like, very few of us routinely experience the stresses of our forebears. With no challenge to overcome, frontier to press, or threat to flee from, the humans of this millennium are overstuffed, overheated, and understimulated. The struggles of us privileged denizens of the developed world—getting a job, funding a retirement, getting kids into a good school, posting the exactly right social media update—pale in comparison to the daily threats of death or deprivation that our ancestors faced. Despite this apparent victory, success over the natural world hasn't made our bodies stronger. Quite the opposite, in fact: Effortless comfort has made us fat, lazy, and increasingly in ill health.

The developed world—and, for that matter, much of the developing world—no longer suffers from diseases of deficiency. Instead we get the diseases of excess. This century has seen an explosion of obesity, diabetes, chronic pain, hypertension, and even a resurgence of gout. Countless millions of people suffer from autoimmune ailments—from arthritis to allergies, and from lupus to Crohn's and Parkinson's disease—where the body literally attacks itself. It is almost as if there are so few external threats to contend with that all our stored energy instead wreaks havoc on our insides.

There is a growing consensus among many scientists and athletes that humans were not built for eternal and effortless homeostasis. Evolution made us seek comfort because comfort was never the norm. Human biology needs stress—not the sort of stress that damages muscle, gets us eaten by a bear, or degrades our physiques—but the sort of environmental and physical oscillations that invigorates our nervous systems. We've been honed over millennia to adapt to an ever-changing environment. Those fluctuations are ingrained in our physiology in countless ways that are, for the most part, unconnected to our conscious minds.

Muscles, organs, nerves, fat tissue, and hormones all respond and change because of input they get from the outside world. Critically, some external signals set off a cascade of physiological responses that skip the conscious parts of our brains and connect to a place that controls a wellspring of hidden physical reactions called collectively fight-or-flight responses. For example, a plunge into ice-cold water not only triggers a number of processes to warm the body, but also tweaks insulin production, tightens the circulatory system, and heightens mental awareness. A person actually has to get uncomfortable and experience that frigid cold if they want to initiate those systems. But who wants to do that? The bulk of us don't see environmental stress in the same light as we do, say, exercise. There doesn't seem to be an obvious reason to leave our shells of environmental bliss.

Maybe that's not entirely fair. In recent years a counterculture has tried to push back against technological overzealousness to reclaim some of our animal nature. They've shucked fancy footwear for flat shoes (and some cases no shoes at all). They've turned away from climate-controlled exercise gyms in favor of rough obstacle courses and boot camps that force muscle groups to work in unison. They're hacking their diets: eating tubers and meat and foregoing grains reminiscent of our Paleolithic ancestors. At least eight million people have bought a product called the Squatty Potty, a device for the toilet to help a person poop in a squatting stance like our pre-toileted forebears did. Millions more sign up for obstacle course races that feature electrified grids, pools of freezing water, and grueling climbs over wooden barriers. They compete until they are so bone tired that their muscles shake. They puke in the mud with tears in their eyes. It's not exhilaration they're seeking: it's suffering. Their pain is so much on the forefront of the experience that the industry of obstacle courses and boot camps are sometimes called "sufferfests." Think about that for a second: There are companies out there that literally make fortunes by selling suffering. How did pain become a luxury good? Could it be that there is a specific sort of pain that might serve a hidden evolutionary function?

It would be wrong to call this movement a fad. To some degree there have always been people who have straddled the line between biology and technology. In ancient Sparta, soldier-scholars wore only simple red cloaks and no shoes, regardless of the weather. They believed exposure made them fiercer in battle and immune to the ravages of the outside world. For almost a thousand years in China and Tibet, mystics and monks endured months or even years on Himalayan peaks with just their robes and daily meditations to protect them. Before Europeans arrived in North America, the natives of what is today the city of Boston wore little more than loin cloths to protect them during the icy winters. In the 1920s in Russia, a movement born from religious fervor convinced hundreds of thousands of Siberians to pour cold water on themselves every day in order to stave off infections and illnesses.

Advanced technology permeates everything we do, but the people who decide to abandon some of that comfort for the rawness of nature represent an indigenous ethos that has almost been wiped out by a societal desire for comfort. They're learning that if they embrace the way their bodies respond to the natural world, they can unlock a hidden wellspring of animal strength.

Today tens of thousands of people are discovering that the environment contains hidden tools for hacking the nervous system. But no matter what they might be able to accomplish, they're not superhuman. The fortitude they find comes from within the body itself. When they forego a few creature comforts and delve more deeply into their own biology they're becoming *more human*. For at least half a century the conventional wisdom about maintaining good physical health has rested on the twin pillars of diet and exercise. While those are no doubt vital, there's an equally important, but completely ignored, third pillar. And what's more? By incorporating environmental training into your daily routine, you will achieve big results in very little time.

It only takes a matter of weeks for the human body to acclimatize to a dazzling array of conditions. Once you arrive at high altitude, your body automatically produces more red blood cells to compensate for lower oxygen saturation. Move to an oppressively hot environment and your body will sweat out fewer salts over time and produce lower volumes of urine. Heat will also stimulate your cardiovascular system to become more efficient and increase evaporation and cooling. Yet no environmental extreme induces as many changes in human physiology as the cold does.

Imagine, if you will, a native Bostonian's experience in the winter. Though beset by ice storms, sleet, blizzards, and constant overcast skies, Boston is not the coldest city in America. But the Boston winters are sufficiently miserable to motivate most of its population to head indoors and jack up the thermostat in the colder months. In Boston, the mean difference between the indoor temperature and the outside air in January is a shiver-inducing 39 degrees. When this typical Bostonian walks out the front door of her stately brownstone she probably cringes with pain as a blast of icy air quickens her nerves and turns her face into a grimace. Beneath the surface of her skin a series of nerve and muscle responses cause the blood vessels to constrict, which can be painful if the underlying muscles haven't been strengthened from repeated prior exposures. If, in a fit of uncharacteristic madness, she decides to remove her shoes and plant her bare feet in the snow, the almost 70-degree swing in temperature would feel akin to walking across a hot bed of coals.

These unhabituated responses of the human body are not pleasant, but the physiology of the process is worth examining. The human circulatory system is made up of a series of spongy arteries and veins that carry our blood supply (and oxygen) to every tissue. Arteries carry red, oxygen-rich blood away from the heart and lungs while blue-tinged veins carry it back. This vast and complex network of vessels would extend more than 60,000 miles if laid end to end. In a single day, the

5.6 liters of blood in a human body travels a total of almost 12,000 miles through the system, or almost four times the distance across the United States. This great blood superhighway is more than just a series of tubes; it's an active and responsive system. Lining most of the important arteries is a similarly complex network of tiny muscles that constrict the flow of blood away from one particular area to boost the supply to another. These muscles are so strong that if someone were to cut off your leg with a sword below the knee, the muscles would immediately clench shut with enough force to almost completely stem the loss of blood. That, luckily, is not the sort of muscular reflex that we need to test on a daily basis, but it's nice to know it's there just in case. However, the second our intrepid Bostonian opens the door to her house and has a brush with that near-Arctic wind, she feels a miniature version of that reaction.

Aside from its lifesaving potential following dismemberment, the circulatory system has other reasons to flex its muscles. To stave off hypothermia, the body conserves heat by shutting off blood supply to the extremities. When this happens, miles of vestibular roadways squinch closed, keeping most of the blood in the body's core and letting the vital organs relax in a warm blood cocoon while temperatures in the hands, feet, ears, and nose plummet. The colder it is outside, the stronger the response. For a person not regularly conditioned to temperature shifts, vasoconstriction is painful. The only way that most of us can trigger the muscle response is to actually go outside and feel the cold. And those of us who live in perpetually climate-controlled environments never exercise this part of our circulatory system.

Weak circulatory muscles are a side effect of living in a very narrow band of temperature variation. The vast majority of humanity today—the entire population that spends the bulk of its time indoors and/or whose only experience when it gets too cold or too hot is wearing state-of-the-art outdoor gear—never exercises this critical system of their body. Even people who appear physically fit, with lean muscles and chiseled abs, might be secretly hiding weak circulatory muscles. And the stakes are huge: In the long run, circulatory diseases contribute to almost 30 percent of the world's mortality.

There's an entire hidden physiology in our bodies that operates on evolutionary programming most of us make no attempt to unlock. Muscle control in the central nervous system breaks down into three distinct categories. There are muscles that we can activate consciously, in what doctors call the somatic nervous system. When you decide to walk across the room, your brain fires the nerves that activate muscles all up through the legs, back, and stomach at once. We don't need to think about every muscle involved in taking a step, we just do it. Still, with some deliberate thought we can individually fire any one of them. This is all part of the somatic system. There are also muscles that we have almost no control over whatsoever. These include muscles that control the pace of the heart, the motion of the vascular system, the speed of digestion, and the dilation of our pupils. All of these are part of the autonomic nervous system—the body's version of autopilot. But there's a third group of muscles and reactions that are shared between the autonomic and somatic systems. Any one of us can decide to take a breath or blink our eyes, but if we let our minds wander, some deep part of our nervous system takes over. If you want to, you can hijack control away from some automatic processes with a thought, but when your mind drifts away, they continue on their own. This is a good thing: With such a system there's no way that you can simply forget to breathe.

The division emerges from deep within our evolutionary roots. Simple life forms respond to the environment in predictable ways. For most mammals, many of these automatic responses originate in the most primitive parts of the brain, near its stem. These relays bypass higher functioning centers in the gray matter. However, as animals encountered more complex and changing environments during the course of evolution, they needed some elements of reasoning to help navigate the world. The cerebral cortex and bigger brain structures, located toward the top of the skull, evolved to accomplish this. Motor functions migrated up into the neocortex, the gray matter areas correlated with higher-level reasoning. Even so, most of the body's millions of actions never go very high in the brain. There has never been evolutionary pressure to put the circulatory system under conscious control, so the response to cold, for example, has been uniform throughout much of our evolution: Preserve the core at the expense of the extremities. No thought needed.

But what happened when humans gained so much technological skill that they effectively dominated their environment? All human technology originates out of the activity of our higher brain functions. In a way, our conscious minds are now so in control of the world we live in that they've left our lizard brains out of the loop. Without external signals and inputs that were designed by evolution over millennia, our bodies are simply not being called upon to perform what have always been critical functions. That internal programming lies dormant and unproductive.

Almost since the beginning of recorded history, humans have seen themselves as separate from the natural world. We divide the planet into two categories: things influenced by human action and things that are untouched. The distinction is false. On a global scale we can see that the constant progress of industry has had a dramatic effect on the climate. The humanizing influence of our carbon footprint affects everything. The year that I'm writing this, 2016, is set to be the hottest ever recorded, expected to top the 10 record-breaking years before it. The scale of the problem indicates that humanity and the environment are intrinsically linked. But does that mean we're making the world more human? Or does it mean that humanity has been part of nature all along?

The tiny muscles around your arteries have one unambiguous answer to that question. Despite everything that we try to do to separate ourselves from the world around us, humans are still indisputably part of nature. As byproducts of evolution, the skyscrapers, plastics, and automobiles we manufacture are no less "natural" than a termite mound, a honeycomb, or a beaver dam. Yes, the actions that humans make may be significantly more destructive or ambitious or awe-inspiring or futile, but they are all part of a greater system of causes and effects. We are still animals. Just very smart ones.

So what does this have to do with the neocortex? Well, if our bodies have preset responses to natural conditions, then maybe it's too simple to think about the limits of the body as stopping at the skin in the first place. Perhaps humans exist in a sort of continuum with the outside world. Our bodies are not discrete things; rather, they are reflections of the environment that they inhabit.

Let me give an example. In the past 40 years, naturalists who study ants have struggled with a similar paradigm shift. There are several distinct types of ants in any given colony. There are worker ants that search out and hunt for food and who carry out the bulk of the manual labor; there are big-headed soldier ants that defend the colony from invaders; and there are breeding ants that constantly churn out new generations. On one level every ant is an individual entity with legs, mandibles, antennae, and an ability to navigate the world on its own. Because you can hold an ant in your hand, dissect it, and analyze its individual parts, it's logical to think of an ant as a single insect. But there is another way to think about that same ant. Instead of millions of distinct insects, today's ant biologists tend to think of the entire colony as a single living organism. When looked at in this way ants are essentially cells of a larger creature. The colony is the body. The group grows in size in the summer and shrinks in the winter. It conquers territory, amasses resources, and gestates a new generation. The sum of all ants together is much greater than the abilities of any individual creature. The colony works as a sort of networked brain: a superorganism.

The body that you have isn't too different from an ant colony. Long before animals ever appeared on earth, in a time when life comprised mostly single-celled organisms, microscopic bean-shaped bacteria called mitochondria flourished in the wild. These single-cell life-forms ate up oxygen from the environment and expelled an energy-rich waste product called adenosine triphosphate, or ATP. Over the course of millions of years, larger single-cell critters needed more energy to perform complex functions. Rather than develop a novel approach to creating ATP, they evolved to absorb mitochondria into their own cellular structures. Thus the first animal cells were born out of a symbiotic relationship. If you were to peer through a microscope into any random red blood cell, you would find thousands of mitochondria sucking up oxygen and excreting ATP.[1] You couldn't survive without them. But that's not all. In addition to mitochondria, scientists estimate you have more than 10 trillion other microbes in your body, comprising more than 10,000 different species, and accounting for 1 to 3 percent of your body weight. Billions more live on your skin, eyes, hair, and in your blood. The realization in recent years that bacteria are vital to human health has spawned the exciting new field of medical investigation into the bacterial genome. Research is showing that the unique mix of bacteria in the human body can have a profound impact on health, and can even create personality changes.

And why shouldn't it? The human genome has 23,000 genes composed of twisted strands of nucleotides called DNA. But there are an additional two million genes in our symbiotic bacterial genome. And, like our own DNA, that bacterial genome gets passed down to our descendants and evolves as we do. In a way, we're actually more microbial than human. Even so, all those different organisms work in concert to create a single human bounded by the barrier of a layer of skin.

That's just our inner space. What happens when we think about the body in terms of its preprogrammed responses to the world? In most cases, the strategies that our bodies used to adapt to stress are completely outside of our conscious control. You don't have to think about sweating when you're working out. Your body just does it. You breathe

---

1   To be fair, the relationship between mitochondria and the human body is a bit more complex than this. Over time mitochondria actually shed some of their own DNA and transmitted it into human DNA, meaning that it would be all but impossible for a present-day mitochondria to live or reproduce outside the body. Still, the mitochondria are genetically distinct from what we think of as human cells.

harder at altitude when you need more oxygen. Your heart and adrenal glands respond to threats before you even have a chance to think about them—giving you extra power in a moment of need. There is an entire hidden world of human biological responses that lies beyond our conscious minds that is intrinsically linked to the environment. The blueprint for these responses is embedded in our DNA and the neural links that we start to develop after we're born. This hidden biology is not part of our higher cognitive functions; rather it is how our unconscious body thinks about the world it inhabits.

For most of our evolutionary past, comfort was a rare treat and stress was a constant. The lower parts of our brain formed in environments where there were always physical challenges to overcome, and those challenges were part of what made us human in the first place. Despite all of our technology, our bodies are just not ready for a world so completely tamed by our desire for comfort. Without stimulation, the responses that were designed to fight environmental challenges don't always lie dormant. Sometimes they turn inward and wreak havoc on our insides. An entire field of medical research on autoimmune diseases suggests they originate from fundamental disconnect between the outside world and an understimulated biology.

This book is largely about what happens when we reexamine our relationship with the environment and see ourselves as a part of something bigger than the comfortable spaces we mostly choose to live in. It explores how changing the environment around the body also fundamentally changes the body itself. More importantly, it shows how it is possible to manipulate our external environment to trigger autonomic responses in predictable ways. Once you realize that you can manipulate deep parts of your physiology by intentionally tweaking identifiable preprogrammed responses, you can begin to cede aspects of that automation to your consciousness.

It's a strange claim to make for an investigative journalist who has spent much of his career trying to debunk false prophets and medical voodoo. For that matter, it's an odd statement for a man whose spirit animal is still mostly made of "jelly." But these findings are grounded in current science and the real lives of people around the globe who have taken control of their bodies to an extraordinary extent.

For me the journey to unlocking my own biology came at a personal low point while I was living in Long Beach, California, back in July 2012. I had been sitting in front of my computer in a desk chair for almost 8 straight hours. Palm trees gently swayed outside my window. Despite my relatively comfortable perch I had a sinking feeling that I was getting past my prime. My legs throbbed from underuse and my back ached. I told myself that as I was now approaching my mid-thirties it was perfectly normal for my stomach to sag over my belt. I thought my days of youthful, vigorous adventure were fading fast, and that this was what middle age was supposed to look like. As an American—hell, as a human—I figured that creature comforts were my best defense against the inevitabilities of growing older. A moderate amount of exercise and an occasional dip into the organic aisle of my grocery store would be sufficient to maintain at least a certain level of decorum.

That was when the internet coughed up a picture of a nearly naked man sitting on a glacier somewhere north of the Arctic Circle. It was as if this bearded gentleman 20 years my senior was thumbing his nose at what had become the defining narrative of my life. His serene blue eyes scanned the landscape and betrayed no fear of freezing to death. It was as if he'd traveled through time from ancient Sparta, when the warriors would pit their bodies against the elements in order to defy the gods themselves. Whatever this guy was into, it wasn't comfort. And yet I couldn't deny that he projected something vital that I'd recently noticed was missing from my own life.

A Google search revealed that his name was Wim Hof, a Dutch guru who claimed to be able to raise and lower his body temperature at will, and control his immune system with only the power of his mind. He ran a training camp in the snowy wilderness of Poland, where people from all around the world converged to study his secrets. He promised that after a short stint, he could teach someone to survive in Arctic environments with almost no gear. He said he had invented a breathing method that strengthened endurance, and that he could teach a meditation routine that would allow anyone a peek into their own hidden biology. What's more? It only took a few days to learn. It all seemed crazy. Hof was clearly another false prophet selling overheated mumbo jumbo and miracle cures. It wasn't like he was the first fake I'd come across. I had just completed almost a year of research on a man who had died while trying to cultivate superpowers under the tutelage of an American-born Tibetan guru in the deserts of Arizona. Ten years earlier, when I was leading a program abroad for college students in India, we did a 10-day silent meditation retreat in a holy place for Buddhism. At the end of the retreat one of my students claimed to be on the cusp of enlightenment, and that realization prompted her to commit suicide. In both cases the search for something greater proved deadly. Those cautionary experiences were the base material for two books I wrote during the first decade of my career. So you could say that I was downright wary of anyone who suggested that humans had hidden powers.

Even so, there was something strangely familiar about Hof's training. I'd seen cognates of it all around Los Angeles. Earlier that month a friend of mine asked me to join along on an obstacle course race that ended with participants shimmying through the mud under electrified wires. I cringed at the idea and declined. Later, during a free yoga class near the beach, I saw a few hundred people costumed as Spartans in full armor charging up a nearby bluff while belting out their signature war cries. And, of course, there were the endless photos on Facebook of friends covered in mud and slogging through hellacious-looking pits of cold water. They never smiled in the photos, but you could tell that they relished overcoming the challenge of it all.

Right then I developed a plan that was as simple as it was cynical. I would travel to Poland to show that my expanding paunch was right all along: The inevitable course of human events was a steady decline into bad health and unhappiness. I was all but certain that Hof was a charlatan who capitalized on the false hope of the gullible masses.

I booked a ticket to Poland with a commission that eventually ran in *Playboy* magazine to try his "method" out for myself. Needless to say, I didn't find what I was looking for. Instead I ended up embarking on my own personal transformation and what is now a 4-year-long journey to become just a little more human.

1

THE ICEMAN COMETH

A DILAPIDATED FARMHOUSE in the Polish countryside creaks and groans on its foundation as six men hyperventilate inside one of its frigid rooms. The windows are caked with frost, and snow piles up outside the front door. Wim Hof surveys his students with stern blue eyes as he counts their breaths. They are lying in sleeping bags and covered in blankets. Every exhalation appears as a tiny puff of mist as the heat of their bodies crystallizes in the near-Arctic air. When the students are bleached white from exhaustion, Hof commands them to let all the air out of their lungs and not to breathe in until their bodies shake and shudder. I exhale into the frigid cold and wonder if it was a good idea to come here in the first place.

"Fainting is okay," he says. "It just means you went deep."

My lungs are empty and my head dizzy from hyperventilation. I note the time on my stopwatch as it slowly ticks forward. At 30 seconds I want to let go and feel the cool air rush inside. But I hold on.

At 60 seconds my diaphragm begins to quiver and I have to rock back and forth to keep from gasping. Even so, my mind is strangely calm. I close my eyes and see red, ghostlike shapes flit behind my lids. Later, Hof will explain that the light is a window into my pituitary gland. I will frown at the explanation. I am the only skeptic in the room.

Hof promises he can teach people to hold their breath for 5 minutes and stay warm without clothes in freezing snow. With a few days of training, he says, I should be able to consciously control my immune system to ramp it up against sicknesses or, if necessary, possibly even use it to suppress autoimmune malfunctions such as rheumatoid arthritis and lupus. It's a tall order, to be sure. The world is full of would-be gurus proffering miracle cures, and Hof's promises sound superhuman. The undertaking resonates with a male clientele willing to wage war on their bodies and pay upward of $2,000 for the privilege of a weeklong program.

Sitting down on a cushion and covered in a sleeping bag nearby, fellow student Hans Spaan's trembling hands are visible. Diagnosed with Parkinson's a decade earlier, the illness forced him to quit his job as an IT executive. Once on the verge of death, he now claims that Hof's method enables him to halve the amount of drugs that his doctors insist he needs to keep his symptoms at bay. Next to him, Andrew Lescelius, a Nebraskan whose asthma can be crippling, hasn't used his inhaler for a week. I've flown halfway across the world to see if Hof is really the miracle worker these people say he is, or if he is just another charlatan offering hope where none is to be had.

We cycle between hyperventilating and holding our breath for almost an hour, and every repetition makes it incrementally easier to hold on just a little bit longer. Hof tells us that the quick breathing adds oxygen to our blood supply so that, at least until we use it up, we won't have to rely on the air in our lungs to survive. The autonomic urge to gasp for air is based on the mind's ordinary programming: No air in the lungs means it's time to breathe. My nervous system has not yet realized there is still oxygen in my bloodstream.[1] The heavy breathing lets me trick my nervous

---

1 Controlled hyperventilation will increase oxygen saturation in the blood to 100%, but more significantly, it also expels CO_2 which your body uses to gauge when to gasp. See page 51.

system into doing things that it isn't evolutionarily designed for. In other words, I am hacking my body.

After 92 seconds, my vision starts to cloud over. The room takes on a red sheen. I might be seeing lights. Another second and I could pass out. I have to let go and feel the air rush into my lungs. It is far from a record, but after only an hour of trying, it is my longest attempt yet. I smile with a small sense of accomplishment.

Hof now commands us to start another breathing cycle, but this time, instead of only holding our breath and sitting in place, we are supposed to do as many pushups as we can. As a 33-year-old journalist without a gym habit who was raised on a diet of processed foods and little exercise, I am out of shape. A week earlier I tested my pushup limit and only managed an embarrassingly feeble 20 before collapsing. But now I've been hyperventilating for almost an hour, and after another short round of heavy breathing I push myself off the floor with almost no effort. The exercises roll out one after another, and before I know it I've tricked—or rather, hacked—my body into doing 40 pushups without taking a breath.

This is the moment that I decide that I will have to reevaluate everything that I thought I knew about gurus. Still, Hof is a difficult figure to dissect. On one level he speaks in a familiar creole of New Age mumbo jumbo. There's a spiel about universal compassion and connection to divine energies. And he has a way of descending into long monologues about how a few simple exercises might foster world peace and, as he says, "win the war on bacteria." After an hour or two of listening to grandiose and even self-congratulatory claims, it is easy enough for me to let my eyes glaze over and ponder the charlatan idea a little more seriously. Then, of course, there are the results. The relatively simple exercises make undeniable changes in my body seemingly overnight. Following his prescriptions for a week straight, I will cajole my muscles into performing feats of endurance that I didn't think were possible, and earn a confidence I didn't know I had. As a bonus, I will lose 7 pounds of fat—which mostly come out in oily clumps during my morning eliminations.

The goal by the end of the week is to complete an arduous 8-hour climb up a nearby powder-covered mountain while wearing nothing but shorts and hiking boots. It will be my own personal Everest, though in this case the mountain is called Snezka. It is the sort of expedition that is an almost hallucinatory bad idea. *Climb a freaking mountain? Nearly naked?* I'd agreed to give the training a chance but being at Hof's mercy seems downright dangerous. And who is he, anyway? When I look up from my string of pushups, I ponder the man wearing a pointy green hat that makes him look like a life-sized garden gnome. A bushy beard frames his piercing blue eyes and ruddy nose, and his body bristles with tightly corded muscles. Hof is nothing less than a savant and a madman; a prophet and a foil. And as is occasionally the case with people who try to cultivate superpowers, Hof's abilities come at a heavy price. A foot-long surgical scar arcs across his stomach and marks a time that his training landed him in an Amsterdam hospital in a fight for his life.

Born in the Dutch city of Sittard in 1959, on the eve of Europe's hippie revolution, Hof spent his early years in the middle of a working-class family of nine children. While the rest of the Hof family learned Catholic liturgy, Wim became fascinated with Eastern teachings, memorizing parts of Patanjali's Yoga Sutras and scouring the Bhagavad Gita and the teachings of Zen Buddhism for wisdom. He was keen on exploring the connections between the body and the mind, but none of what he read was quite what he was looking for.

Then, in the winter of 1979, when he was 20 years old, he found it. While walking alone on a frosty morning in Amsterdam's picturesque Beatrixpark, he noticed a thin skin of ice on one of the canals. He wondered what it would feel like if he jumped in. With a juvenile impulsiveness he has never quite shed, he took off his clothes and plunged in naked. The shock was immediate, he says, but "the feeling wasn't of cold; it was something like tremendous good. I was in the water only a minute, but time just slowed down. It felt like ages." A wash of endorphins cruised through his system, and the high lasted through the afternoon. He has gone on to repeat the exercise just about every day since. "The cold is my teacher," he says.

The breathing technique emerged naturally. He started by mimicking the rapid breaths people take instinctively when they plunge into icy water, which he says are similar to the breaths a woman takes during childbirth. In both cases the body switches to an instinctual program. When Hof dunked under the ice, his body went naturally from rapid breathing to a breath hold. That's when he began to feel changes in his body.

The way Hof explains it, humans must have evolved with an innate ability to resist the elements. Our remote ancestors marched across endless expanses of frosty mountains and navigated parched deserts long before they invented the most basic footwear or animal-skin coats. While technology has made us more comfortable, the underlying biology is still there, and Hof believes the key to unlocking our lost potential lies in re-creating the sorts of harsh experiences our ancestors would have faced.

Hof trained on his own in obscurity for 15 years, rarely talking about his growing abilities. His first student was his oldest son, Enahm. When Enahm was still an infant, Hof took him down to the canals and dunked him in the water like Achilles. While it's anyone's guess what nearby pedestrians might have thought of this sight, most of his close friends shrugged off his morning routines as just another eccentricity in an already eccentric city.

Hof did odd jobs, including working as a mail carrier, and took summer gigs in Spain as a canyoneering instructor. Money was always a problem, and his wife—a beautiful Basque woman named Olaya—began to show signs of a serious mental disorder. She was depressed. She heard voices. In July 1995, she jumped off the eighth floor of her parents' apartment building in Pamplona on the first day of the Running of the Bulls.

Sitting at a handmade wooden table in what serves as a lunchroom and breakfast nook in his Polish headquarters, Hof recounts Olaya's death as tears roll freely down his cheeks. "Why would God take my wife from me?" he asks. Confronted with loss and a broken heart, he put all his faith into the one thing that set him apart from everyone else: his ability to control his body. Olaya had never shown interest in Hof's methods, but he felt he could have done more to help her. "The inclination I have to train people now is because of my wife's death," he says. "I can bring people back to tranquility. Schizophrenia and multiple-personality disorder draw away people's energy. My method can give them back control." It was his call to action. But he still needed a way to announce himself to the world.

His opportunity came a few years later. As winter settled on Amsterdam, a local newspaper ran a series of articles about various odd things people did in the snow. The paper contacted Hof and he explained that for the past couple of decades he'd been skinny-dipping in icy water. The paper sent a reporter, and Hof jumped into a nearby lake he frequented. The article ran and the next week a television crew showed up.

In that now-famous segment, Hof cut holes in the ice and jumped in while the news crew filmed. He was drying himself off when, a few meters away, a man stepped on a thin patch and fell through. Hof charged out onto the lake, jumped in a second time, and dragged the man to safety. The news crew caught the exchange, and soon Hof wasn't just a local oddity, he was a local hero. Someone dubbed him the Iceman, and the name stuck.

After that act of heroism, Hof became a household name across the Netherlands. A Dutch television program hosted by the eminent newscaster Willibrord Frequin asked Hof to perform on camera. The gimmick was to have Hof establish a Guinness world record. They planned for him to swim 50 meters beneath Arctic ice without breathing. It would be sensationalist fun, but the program would air throughout northern Europe and give Hof a shot at doing stunts for other channels around the world.

A few weeks later Hof stood on the surface of a frozen lake near the small village of Pello, Finland, a handful of miles from the Arctic Circle, wearing only a bathing suit. Although the temperature would drop to minus 12 degrees Fahrenheit, his skin glistened with sweat. Below him a diamond-shaped hole shot down a meter through the ice. There were two other holes 25 and 50 meters from the first. A camera crew watched as Hof descended and dipped his toe into the periwinkle waters.

On the first day of shooting he was supposed to swim only to the first hole so that the crew could set up the right shots and feel comfortable with safety protocols. But Hof had other plans. He wanted to surprise and impress the crew by clearing the whole distance in one go. He had done his calculations in advance. One stroke took him just over a meter and he calculated that he would need to do 42 to reach his destination. Taking a giant gulp of air into his lungs, Hof disappeared and began his sprint.

He later recalled that he opened his eyes midway between the first and second hole and could make out a beam of sunlight slicing through the water. But at stroke 29, with the safety of the first hole and rescue team behind him, something went wrong. He hadn't anticipated what the cold water would do to his eyes. His corneas began to freeze over, and crystallization blurred his vision. Five strokes later he was blinded, with only his stroke count to direct him to oxygen. Soon he was off course. At 42 strokes he grabbed around in vain for the rim of the second hole. He turned around thinking maybe he had passed it. He wanted to gasp for air but knew the results would be fatal. At 48 strokes his hope was beginning to fade. Seventy strokes in, just as he began to lose consciousness, he felt a hand wrap around his ankle. A safety diver dragged him to the surface. He knew he had almost died and that his hubris had led him to the brink. Despite that close call, the next day he would set a world record, with the cameras rolling.

The show went on to be a hit and secured Hof a series of similar on-air stunts for international channels from Discovery to National Geographic to Vice. They seemed to grow in proportion to one another—with each one pushing death's barrier. There was a barefoot marathon above the Arctic Circle, where the ice numbed his feet so thoroughly that his toe turned black with frostbite. The injury healed on its own despite a doctor's assessment to the contrary. When he aimed to head up Mount Everest a few months later, again with a film crew behind him, he started his ascent to worldwide media attention. His attempt to reach the summit in his shorts ground to a halt at 25,590 feet, roughly 3 hours from the top of the earth—just inside the so-called death zone that has taken the lives of at least 250 climbers—when his foot started to experience a familiar numbness that he'd felt in the Arctic. Rather than risk losing his foot he turned back. Success could mean too high of a price. Although he was capable of incredible feats, Hof's desire to impress and please the people around him time and again had led him into near-fatal situations. Should he die, the world might never understand how he had achieved his dramatic results. Hof needed a better plan.

MY RELATIONSHIP WITH Hof starts in the winter of 2013 when I climb on board a plane from Los Angeles to Wroclaw, Poland, with a magazine commission and the intention to debunk his claims. At best, I believe Hof is a circus act who has mastered a few tricks of endurance and sleight of hand. He won't be the first false guru that I'd disrobed in print. Just a few months earlier *Playboy* published an article of mine about a monk in Arizona named Geshe Michael Roach, who was teaching a bastardized form of Tibetan Buddhism that allowed him to have free range over the sexual activities of some members of his flock. Roach promised his students that they could meditate their way into superpowers. With his own brand of training they could learn to become invisible, read people's minds, and transform into enlightened angels. That might have been unremarkable in a world of eccentric ideologies, except that the spiritual teachings of this "lama"—a Tibetan title given to master practitioners—led a young, idealistic follower to meditate until he died of dehydration in the mountains above the cultic compound. The week I set off to meet Hof is also the one in which I secure a deal to expand the article about Roach and that sordid affair into a full-fledged book that I intend to be my opus against sham spirituality. To me, Hof seems that he is trying to use his history of carnivalesque stunts to amass gullible followers to his "method" and empty their wallets in the process. In other words, I perceive my visit with Hof to be a mere pit stop on my career as an investigative journalist debunking people like him.

Of course there are some formalities to get through first. The noblest ideals of journalism include hearing all sides to any story and giving sources a fair shake. If Hof claims to have a method to consciously control parts of his body that are locked away from most people, then I can't simply write from the sidelines. Instead, I must follow his instructions and watch the claims fall apart. Over the years I've interviewed organ traffickers, interrogated mob bosses, and been held up by child soldiers with automatic weapons. As long as I don't plunge fatally into hypothermia, this isn't the most dangerous assignment I've ever accepted.

After my plane thunders into the terminal gate, Hof meets me on the other side of baggage claim with a broad smile. He made his headquarters in Poland instead of his home in the Netherlands so that he can be closer to icy streams and snow-covered mountains—and also to take advantage of a weaker economy that allowed him to purchase a larger space. We pile into a tiny gray Opel Astra with two other devotees—a Croatian and a Latvian—who've also come to study his technique. Together we traverse miles of Polish pines and picturesque villages en route to Hof's rural base of operations.

Janis Kuze crams next to me, and my hiking backpack overflows onto his lap. The burly Latvian grew up amid the turmoil of a collapsing Soviet Union, when bandits roamed the countryside. His father stashed a loaded AK-47 beneath his son's bed so that it was never more than an instant away should they have needed to defend themselves. Now Kuze is studying the Israeli martial art Krav Maga in his spare time and spars with his equally intimidating, and he assures me beautiful, girlfriend. Asked if he was ready to immerse himself in ice water, he replies, "When my father was in the special forces, they tested soldiers' ability to adapt by making them sit in ice water. If they survived, they passed. Not everyone passed." The Croation, Vladamir Stojakovic, works for a mobile phone company back home and is looking for something that might reinvigorate his office-bound life.

We arrive in the tiny village of Przesieka, where Hof owns the isolated farmhouse he was able to purchase after signing a deal in 2011 with Columbia Sportswear to shill a line of battery-heated jackets. In the commercials, which were created for TV but thrived on the internet, Hof swims in a frozen lake while giving icy stares to toasty outdoorsmen who use the high-tech gear to warm themselves at the touch of a button. The videos went viral, and commenters compared Hof to Chuck Norris, propelling him to a sort of internet alpha-male celebrity. But the condition of the house confirms that fame does not always translate to riches. The space is a permanent work in progress, with an assortment of bunk beds and yoga mats strewn pell-mell across the two floors. A busted sauna sits next door to its newer replacement, and a coal furnace that doesn't quite work spews black smoke through cracks in the floorboards. For that matter, most of the floors themselves don't seem level.

Still, the crumbling building is the headquarters for Hof's growing global presence as a New Age guru and ground zero for the experimental training regimen he is developing. One of Hof's first students at the house was a college student named Justin Rosales, who flew to the Netherlands from Pennsylvania in 2010 to serve as a guinea pig. "If we want to become strong, passionate, and motivated, we have to take on seemingly impossible tasks. Without an open mind, the cold will never be your friend," Rosales told me over e-mail. He wrote and self-published a book with Hof about his experience, entitled *Becoming the Iceman*, which is often passed around among devotees looking for personal transformation.

I stash what little winter gear I'd brought beneath a bunk on the second floor and look out the window onto a snowy field that serves as the main training site. What I see astonishes me. I watch Andrew Lescelius, the wiry, asthmatic Nebraskan who arrived a week earlier, cross the space clad only in black underwear. He stops to pick up handfuls of snow and then commences to rub them over his arms and chest. Steam erupts off his body in great clouds. My mind tries to grasp what I am looking at. I am simply unwilling to comprehend the unfolding scene. *Is that actual steam?* Kuze, who chooses a bunk next to mine, looks out at Lescelius. He tells me that he is eager to get out into the snow. Somehow the idea of getting naked in the ice excites, rather than dissuades, him. I let him head outside on his own. I will have plenty of opportunities to be cold in the coming week.

After a restless night of sleep, I take my place with the group that gathers in the yoga studio downstairs to meet Hof in our first official lesson. The program is still clearly in its infancy, and Hof explains that every training session will be a little different depending on the chemistry of the group. But no matter how it starts, the building blocks are simple enough and, he assures us, our progress will be rapid. "This week we will win the war on bacteria!" he proclaims (again) before warning us that he will challenge everything we ever thought about the limits of our bodies. The monologue goes on a touch too long. There's a tendency Hof exhibits in many of his public appearances to be heavy on promises but often light on practice. But eventually he tells us to shed our clothing and head outside.

We exit the farmhouse in our underwear and bare feet and trod onto the snowy field frequented by deer. I notice the curious gazes of inquisitive neighbors. As we file past, one of them yells something in Polish and Hof chuckles. Most people here think he's crazy, if affable. Not that I am paying too much attention to the opinions of fully clothed Polishmen. It is the first time in my life that I've put my feet directly into snow and they feel as sensitive as a newly broken tooth. My heart rate jumps. Kuze lets out a gasp while Hof beams a trickster's smile. We stand in a circle and take low horse stances—a wide-legged posture derived from Asian martial arts that resembles the position of a person riding a horse. The goal is simple: to focus our minds on our foreheads and simply endure the cold with chests bare to the air. Five minutes is excruciating, but Hof has us stand for 6 before sending us numbly to the sauna.

With numb limbs, going from a snowy field into a 100-plus-degree sauna feels like a sensational mistake. The pain is even worse than standing in the snow, something that I could not have imagined was possible. The body's natural reaction to cold is self-preservation. To keep the core warm, the muscles that control the arteries clench tightly and restrict the flow of blood only to vital areas. The process is known as vasoconstriction. This is why frostbite starts in the extremities: the lack of blood flow to those areas makes them cool much faster than if they were flush with warm blood. The sudden change to heat has the opposite effect. Arteries suddenly pop open and blood surges back into those cold areas, generating an excruciating wave of pain.

Kuze stretches his feet toward a box of coals and confides that he is on the verge of tears. Lescelius clenches his teeth and holds his breath. A side effect of asthma, he tells me, is poor circulation, and the sensation of vasoconstriction is even more painful for him than it is for me. "But I like to think of it as lifting weights for the circulatory system," he says. Hof nods at the statement. We aren't just warming our bodies, we are feeling new sensations. It is an early step to gaining control over them. After years of exposing himself to the cold, Hof says that he can now operate his arteries much like he could his fingers. That is, he can consciously restrict the blood flow of his limbs and send it to any part of his body that he wants to.

Although the first day of exercises is painful and exhausting, true to Hof's word our progress is rapid. The next day we stand in the snow for 15 minutes before the same feeling of panic sets in. In the afternoon we take a brief dip in the basin of an ice-cold waterfall that is a 5-minute walk from Hof's back door. With every attempt, the barriers that we've built in our minds about the cold recede a little further.

By the fourth day, standing in the snow is barely a challenge. An hour passes by quicker than 5 minutes had just days earlier. In the evening we sit on snow-covered rocks by a stream until they are warm. All the while Hof smiles over us like a Cheshire cat.

What we know about how the human body reacts to cold comes mostly from gruesomely accurate studies that emerged from the Dachau death camp. Nazis tracked Jewish prisoners' core temperatures as they died in ice water. As terrible as they are, these morally compromised studies have helped doctors understand how quickly the body loses heat in such conditions. Sitting in 32-degree water, humans begin to feel sluggish after only a minute or two. By 15 minutes most people fall unconscious. They die between 15 and 45 minutes depending on their underlying physiology. When the core body temperature falls below 82 degrees, death is almost inevitable. Measured against that data set, Hof seems to perform miracles.

In 2007, at the Feinstein Institute for Medical Research on Long Island, Kenneth Kamler, a world-renowned expedition doctor who has worked on Everest, observed an experiment in which Hof was connected to heart and blood monitors and immersed in ice. At first the experiment hit a major snag. The standard hospital devices that track respiration declared him dead after he'd been in the ice only 2 minutes. The machine got confused because he hadn't taken a breath and his resting heart rate was a mere 35 beats per minute. He wasn't dead, though, and Kamler had to disconnect the device to continue. Hof stayed in the ice for 72 minutes. The results were astounding. Hof's core temperature initially declined a few degrees but then rose again. It was the first scientific validation of Hof's method. "Exactly how you explain it depends on the kind of philosophy you want to believe in," says Kamler when I later track him down on the telephone. He references similar feats called *tummo* performed by Tibetan monks where they modulate their body temperature with breathing and intense visualizations of tantric deities. Ultimately, he says, it boils down to how Hof uses his brain. "The brain uses a lot of energy on higher functions that are not essential to survival. By focusing his mind, he can channel that energy to generate body heat," he speculates.

Interest in Hof among scientists snowballed in 2008, just as it had in the mass media about a decade earlier. Researchers at Maastricht University in the Netherlands wondered if Hof's abilities stemmed from a high concentration of mitochondria-rich brown adipose tissue, also known as brown fat. This little-understood tissue can rapidly heat the body when it metabolizes its fuel source: ordinary white fat. Brown fat is what allows infants—who don't have the muscles to warm themselves like most adults—not to succumb to cold in their earliest months. Usually brown fat mostly disappears by early childhood, but evolutionary biologists believe that early humans may have carried higher concentrations of it to resist extreme environments. The scientists learned that Hof, then 51, had built up so much brown fat over the course of his training that he could produce five times more heat energy than the typical 20-year-old—most likely because he repeatedly exposed himself to cold.

Brown fat may be the missing organic structure that separates humans from the natural world. White fat stores caloric energy from food, which the body tends to burn only as a last resort. In fact, it's difficult to burn the spare tire off your waistline because the body is programmed to store energy; even during intensive exercise routines the body will burn muscle before it uses white fat for energy. Brown fat is different. Most people create it automatically when they're in cold environments by way of a process called beiging. Essentially, the body detects physical extremes and starts to store mitochondria. When brown fat starts to work, the mitochondria suck white fat through the bloodstream and metabolize it directly to generate heat. Because most people do everything they can to avoid environmental extremes, they never build up brown fat at all. If we lived without clothing the way our distant ancestors must have, then we would have to rely on the internal properties of brown fat to keep us alive during the winter months and cold nights.

As we sit in the sauna, I ask Hof how someone activates brown fat consciously. Instead of explaining, he launches into a demonstration.

He clenches the muscles in his body in sequence, from his toes to his rectum then from there up to his shoulders, as if pushing something up from below. Then he furrows his brow and squinches down his neck as though he is trapping that energy in a point right behind his ear. The process turns his skin bright red, as if he is catching fire. Suddenly he kicks out his leg and falls against the wall gasping. "Oh my God," he says, dazed. In his eagerness to teach, he didn't factor in the heat of the sauna and almost blew a fuse. He lurches out of the sauna and rolls in the snow outside. The few of us pupils exchange concerned glances, but in about 15 minutes he returns with an embarrassed smirk. "That's how you do it. But try it only in the cold."

Hans Spaans, who was diagnosed with Parkinson's disease in 2004, credits Hof with saving his life. When I first met Hans in Poland, I discovered that he has been trying the techniques for a little less than a year, mostly in private sessions with Hof. "With this disease," he says, "most people have to take more and more drugs just to maintain the same level of mobility and quality of life, and eventually you max out and begin the long decline." Spaans is trying to manage his drug regime by accompanying it with Hof's breathing technique and ice-cold showers. He tracks his drug use on spreadsheets and claims to be on far fewer drugs now than when he was first diagnosed. He credits Hof with keeping him out of a wheelchair. Although such anecdotal evidence is encouraging, it's hard to determine the extent to which the benefits of Hof's method can be chalked up to the placebo effect. Since Hof claims to be able to control his own autonomic nervous system—one of the systems affected by Parkinson's—it is important to verify these claims through scientific study.

Peter Pickkers is just about the last scientist who would be swayed by outlandish claims. A research scientist at Radboud University Medical Center in the Netherlands and expert on sepsis and infection, he specializes in the human immune system. In 2011, Hof met Pickkers and mentioned that he could suppress or ramp up his immune system at will—a feat that is, by definition, almost impossible. But Pickkers, who had watched Hof's career rise on TV, was curious.

Pickkers, along with his protégé—a graduate student Matthijs Kox, who would go on to use this as the basis for his PhD thesis—devised a test in which he administered endotoxin, a component of *E. coli* bacteria. The body thinks the substance is dangerous, but it is actually inert.

Pickkers's pioneering work proved that 99 percent of healthy people who come in contact with endotoxin react as though they have the flu before the body realizes it has been duped and returns to normal. The test has been useful to figure out the effectiveness of certain immune-suppressant drugs, which are the sorts of pharmaceuticals an organ transplant patient might take to stop the foreign tissue from being rejected. If a prospective drug stops patients from reacting to endotoxin, then it is a good indicator that the drug worked and is effective in turning off their immune systems.

While Hof meditated, Pickkers and Kox injected him with the endotoxin. The results were unheard of. "Wim had done things that, if you had asked me prior to the experiment, I would not have thought possible," Pickkers tells me. Whereas almost every other person dosed with endotoxin experienced severe side effects, Hof had nothing more than a minor headache. Blood tests showed he had much higher levels of cortisol—a hormone usually released only during times of extreme stress, sort of like adrenaline—than Pickkers had recorded in any previous study. Also, blood drawn while he was meditating remained resistant to endotoxin for 6 days after it had left his body.

Hof is unambiguous when I ask him about the results: "If I can show that I can consciously affect my immune system, we will have to rewrite all the medical books," he says. But Pickkers and much of the rest of the scientific community are more reserved. While the results show an unprecedented response to endotoxin, it was maybe more likely that Hof is just a genetic anomaly—the 1 percent of freaks who for some reason just don't react to endotoxin. Still, the results were promising enough for Pickkers and Kox to commission a second study. In the upcoming experiment Hof will guide a group of college students through the same basic course I am taking now before heading back to Pickkers and Kox's lab for a good old-fashioned scientific prodding. If his technique proves to be teachable, then the ground could shift under Pickkers's feet. The experiment will take place the week after I complete my own training. And I will have to wait more than a year to learn the results.

For now I will have to focus on a lone challenge I have yet to complete, one that is just a little more daunting than a dip into a waterfall or lying in the snow. Despite my progress, I am not sure I will be up for the grueling bare-chested hike straight up the mountain. Mount Snezka straddles the Polish-Czech border and is battered by icy winds throughout the winter months. At its 5,259-foot summit, frequented mostly by intrepid cross-country skiers who hike up from a ski lift, a lonely observatory records the movements of the stars. We set out early in the morning, just as the sun sheds a few rays of light up into the sky. Starting at the base of the mountain, Hof, three other disciples, and I begin the arduous climb through 2 feet of fresh powder. Seconds after we pile out of Hof's dilapidated Volkswagen van, the cold slices through our winter coats. At 25 degrees Fahrenheit the modest breezes feel excruciating. In the parking lot, skiers clad head to toe in colorful Gore-Tex ensembles wrestle with their gear and trek slowly to the chairlifts.

Hof leads us to a side trail that snakes through parkland to the summit. Ten minutes up the trail, after our bodies have had time to build some internal heat, we strip off layers. Ashley Johnson, a former English hooligan who found new direction in life doing volunteer work around Hof's house in exchange for lessons, slaps Lescelius and Kuze on the back in camaraderie. Bare to the cold, we stash our clothes in our backpacks and crunch forward through the powder.

The moment I take off my shirt I start to get a sense of how our primordial ancestors would have handled the same march. Trudging forward I don't feel the bite of the cold the way I did before undergoing the regime at camp. Whatever heat I build up through exertion seems to stay in my skin as if I am wearing a wetsuit. I can feel the cold, but it never penetrates. I never shiver. Instead, I focus on the point behind my ears that Hof says will help activate my brown fat and send waves of heat through my body.

I imitate what I witnessed Hof do in the sauna. With my muscles clenched and my mind focused, it isn't long before I start to sweat. A thin, steamy mist wafts upward from the group. A skier stops to take pictures. A ski patrolman on a snowmobile pauses to see if we were okay. A snowboarder lets out a shocked cry and speeds by as we plod forward to the summit.

The strangest thing is that we aren't suffering at all; we are elated. Cold and exercise trigger an enormous endorphin release that pastes a maniacal grin across my face. It is as if the temperature is subservient to the task ahead. Six hours later I near the summit, bare-chested and with my legs caked in snow. In just a week I've gone from California palm trees to Poland's snowy peaks and I feel perfectly warm—hot, even.

But it isn't over. The mountain is steep, and despite our effort the progress is slow. Every step upward through the tree line leaves us incrementally more exposed than before. By afternoon the outside temperature drops to just 8 degrees Fahrenheit, and at about 300 feet shy of the summit, something changes. My core temperature is fine, but the wind intensifies and the incline grows steeper. Every step feels harder than the one before. I am fatigued, and I begin to regret my lackluster gym habit. I worry about what might happen to me if I stop. Would the cold break through the mental barrier I've erected and send me cascading into hypothermia? Fear, more than anything else, keeps me walking. It isn't the time to disprove the method. Twenty minutes later I reach the summit, not cold but more tired than I can ever remember being before. We take a couple of photos, and then I roll in the snow and marvel at how warm I feel. Then we step into the observatory together.

Just like when I walked into the sauna after standing in the snowy field that first day of camp, the warm air of central heating sends a shock through my system. A minute earlier I felt immune to the elements and happily cavorted in the snow, but now that I am out of harm's way and have begun to relax I feel a new creeping, cold sensation. My mental armor is gone and the ice in my extremities leaks into my bloodstream. Warm environments are supposed to make me feel warm, but that's not what is happening. Things are backward. At first a small shiver escapes up my spine. A minute or two later it is a full blown shake. My teeth clatter audibly and soon I can't remember ever being this cold before in my entire life. This, I will learn later, is what paramedics who rescue hypothermic patients call "afterdrop." During vasoconstriction the limbs get much colder than the core. When the body starts to warm up, it lets blood through the vascular system again. Blood chills as it passes through the cold arms, hands, and feet, and when it eventually circulates back to the heart it lowers the body's overall core temperature. In extreme cases, afterdrop can be fatal. An hour elapses before I feel ready to get back on my feet for the climb down the mountain. No longer focused enough to let my body resist the elements, I don the black peacoat that I brought up in a backpack.

On the way back down Hof tells me that he wants to try to summit Mount Everest again. It will be his second time after his earlier, aborted, nearly naked attempt. I ask Hof what he thinks will happen if he finally reaches his limits on that climb and joins the hundreds who have died on the mountain. Would his message be lost to time? Would even the modest lessons he has been able to give to his flock mean anything if he dies in a way most people would deem foolish? His face grows dark at the thought. He looks like he might cry. "I must not die," he says. "I've decided."

The next day Hof brings me back to the airport in Wroclaw. I feel a certain sense of accomplishment. The cold doesn't scare me anymore. In just one week I had dropped almost 7 pounds of pure fat—a welcome side effect of repeatedly warming up my body. But what exactly had happened on Mount Snezka? Had I extended my abilities that had been locked away in an evolutionary dustbin? Or had I only lurched perilously close to the precipice of hypothermia? In other words, was I stronger, or did I just *think* I was stronger? Had I merely delayed feeling cold until I started to warm up, only to put myself at serious risk from afterdrop? To answer these questions, it occurs to me that I will have to understand a little bit more about how we became human in the first place.

2

RAIDING EVOLUTION'S DUSTBIN

IN 1931, ARTHUR FOX, a chemist at the headquarters of the chemical company DuPont, was working in a laboratory with his partner, C. R. Noller, trying to make sense of a chemical that had the peculiar ability to turn fish skin transparent. The finely milled white powder was called phenylthiocarbamide, or PTC by people who find seven syllable words to be a mouthful. As Fox grabbed for the container of PTC, it slipped from his grasp and exploded onto the floor, filling the room with a giant white cloud. Chemical dust covered Fox. It coated his clothes and filled his mouth. In this moment he was probably wondering if he might start to turn slightly translucent. Just a few feet away from him, his colleague, who was also caked in powder, began to complain of an intensely bitter taste in his mouth. It was awful, he said, but Fox tasted nothing. Curious about the disparity in sensation, and apparently unfazed by this dangerous breach in safety protocol, Fox dipped his finger into the powder and deposited another glop on his tongue. Nothing. Noller tried a second dab and grimaced at the bitter taste. This humble moment, which could have turned into a running practical joke between the two men, instead sparked a flurry of research into discrepancies in the ways that humans experience the world. Some of us, like Fox, can taste a normal array of things. But then there is a certain percentage of supertasters like Noller for whom Brussels sprouts are obnoxious, lemons pucker-worthy, and salt more profoundly, well, salty. In the course of their subsequent investigation they discovered that the difference in their taste sensitivity derived from an abnormally high number of mushroom-shaped taste buds on Noller's tongue. People like him, these supertasters, account for about 25 percent of the population. And, for whatever reason, women are more likely to be supertasters than men.

While supertasting does not have much of a profound effect on most people's daily lives, the existence of supertasters offers a peek into to the strange world of sensory perception. Indeed, PTC sensitivity is far from the only recently discovered sense. There are people who can see more than 100 million colors (the normal human can distinguish about 2.4 million), an ability called tetrachromacy. There are people who have the ability of absolute pitch and are able to listen to a note on a piano and instantly recognize its frequency without needing another tone for reference. There are blind people who can navigate through the world by clicking their tongues and listening to the echoes—an ability that they share with bats and dolphins. Some people's sensory hardware scrambles the inputs between senses so that they can taste colors and hear smells, a condition called synesthesia. The neurological jumble usually doesn't interfere with synesthetes' daily lives, and indeed many go on to be extraordinarily successful and creative—from the author Vladimir Nabokov to the scientist Richard Feynman.

While these superabilities don't always have an immediate application to life in the world we live in now, they offer underlying opportunities for evolution to take advantage of the mutations. Should the occasion present itself there may be situations where tasting a wider array of chemicals or seeing more of the visual spectrum could affect a person's chances to reproduce and pass on their genes.

Some of these abilities, like absolute pitch, can even be trained such that those possessing the gift can learn to pick out the notes of a song instantly, as if they were reading them on paper. At birth someone who has the mutation for absolute pitch might indeed have the raw potential to become an expert musician, but only with training and concentration can that individual hone the sense for optimal performance. Once trained, that inborn ability becomes "perfect pitch," a skill and ability prized by musicians. But if left untrained, musical ability can disappear with disuse like a vestigial tail, even when the underlying biology for absolute pitch is still present. A person with such a gift but with no musical inclinations could live their whole life without ever realizing the ability they might have had.

Indeed, it doesn't take much effort to find some abilities that our entire species is in the process of pushing to the wayside. A century and a half ago the darkest corners of the world weren't only being explored, they were being exposed to the unstoppable progress of a global economic engine. After explorers reached the North Pole, the people after them brought general stores to the Arctic. Highways bisected impenetrable jungles, and air travel made every part of the planet reachable within hours. Today, getting from one remote island in the central Pacific to another is literally as easy as booking a ticket online.

Before all that, however, the world was incomprehensively vast and unknowable to all but the greatest explorers. In the islands of the Pacific, solitary men grappled with the unfathomable distances of water around them, yet they devised methods to navigate their way between islands in tiny canoes with only the stars and stories for guides. Without maps, Pacific Islanders paddled between habitable polka dots of land on journeys that could take weeks. Some anthropologists believe that they might have been able to cross the endless open water of the Pacific all the way to South America using nothing more than their own dead reckoning—possibly setting up small settlements in the Americas 10,000 years ago. While knowledge of the stars can orient sailors at night, weather conditions in the Pacific are rarely predictable enough to allow for the cloudless sky views needed to rely on that method alone. Indeed, Captain Cook—the most famous English explorer of the Pacific and the first European to circumnavigate New Zealand—achieved much of his success with the aid of a Tahitian chief who boarded his ship, the *Endeavour*. The chief, named Tupaia, sailed with the crew for almost 20 months and used his memories of the waters to help Cook create a map of 130 islands along a 2,500-mile swath of ocean. Much to Cook's astonishment, no matter where the *Endeavor* traveled, and despite turbulent conditions, dark of night, or the stillness of a featureless day, Tupaia could always point to the direction of his home island. Tupaia was a wave pilot who could read the minute signatures that land leaves on ocean currents, something he called *di lep*. Cook, who had the aid of a Royal Navy compass, recorded in his account of his trip that Tupaia was never once wrong with his directional sense.

Back in the time of the *Endeavour*, Captain Cook landed on the northern tip of Queensland in Australia. There he found an aboriginal group called the Guugu Yimithirr, and decided to create a codex of their language. Among the words he wrote down were "kangaroo" and their word for ocean. He remarked at how different the language sounded to his ear than every other language he had heard in the Pacific. Most interestingly, the Guugu Yimithirr had such a developed sense of direction that when they described the location of things it was always in terms of the cardinal directions. They had no words for "right," "left," "front," or "back." If you sat down next to a Guugu Yimithirr tribesman at dinner, you wouldn't be on their right side, you'd be to the West, or East, or North, or South. This undoubtedly made Cook's cross-cultural dinner party a confusing affair, if also an interesting anthropological moment. Despite not putting themselves at the center of their language, these Aborigines could talk and communicate their orientation even if they were in an enclosed room, or without lights. Knowing where they were was simply second nature. Today, the anthropological research on their language is all but extinct. The tribespeople mostly scattered to the wind, or became the victims of centuries of persecution.

Tupaia, and many indigenous people for that matter, had an innate sense of special orientation that is all but nonexistent in modern society. But there is reason to believe that this latent sense—termed "wayfaring" by anthropologists—is something that many people still have biologically hardwired. In the 1970s a biologist at the University of Manchester in England named Robin Baker attempted to test a theory that some humans, like birds, are able to sense direction through the use of an innate cellular compass. The theory was that humans have magnetically sensitive cells in their nasal bones and eyes that can feel the pull of the magnet poles in the same way that compass needles always point north. Baker decided to test the idea by tying magnets to the heads of healthy college male volunteers in order to reorient any magnetic cells to the local interference. He tied nonmagnetic brass bars to the heads of his control group and then drove his blindfolded test subjects into the English wilderness and asked them to point the way home. In his groundbreaking findings, he reported that the nonmagnetized students could point the way home with a significantly higher degree of accuracy than the others. The research set off a flurry of subsequent research, not all of which was able to replicate Baker's results. Other magnet-tying tests failed. However, after many contentious debates carried out over the course of a decade, researchers reached a consensus that whether the cause was magnetism or some other aspect of underlying biology, humans do indeed have an innate directional sense. It's just that we don't use it very much anymore.

For anyone whose lifespan encompasses the transition from the analog 1990s into the hyperconnected 2000s, the loss of innate wayfaring is easy to recognize. In my teens and twenties the only way to learn a new city was to get familiar with maps. One of my first jobs ever was as a door-to-door fundraiser for an environmental group that promised good money if I could convince householders to cough up cash for a campaign to shut down polluting power plants around Boston. The summer job spoke both to my perpetually empty wallet and my latent desire to do something good for the environment. After a few weeks I rose to the position of field manager, which meant I was in charge of coordinating a summer's worth of door-knocking for a crew of canvassers throughout the greater Boston area. Budgets and time were tight, and the crews had to work as efficiently as possible. I color-coded a paper map of my patch of the city: the suburb of Somerville. It took just a few days before I knew every alley, dead end, and thoroughfare of the city's incomprehensible web of streets. Constantly reviewing the spatial information in front of me as well as guessing likely routes between pick-up and drop-off points helped me internalize the entire city. Name a street and I knew precisely how to get there. And then, a few years later, came the digital revolution.

Sometime in late 2009, around the same time I moved from Brooklyn, New York, to Long Beach, California, I bought a digital map that suction-cupped to the windshield of my car. Up until that moment I'd missed the smartphone sea change, and I was entranced by how this little device allowed me to enter an address and follow simple, in-the-moment directions to plan my way across my new city. The paper map that had brought me on a road trip across the country gathered dust in my trunk, and I used the TomTom every time I got in the car. Every turn was a carefully choreographed digital prompt. Six months later I could barely make it across town without it. My total dependence on the device led me on some startlingly ill-advised wild goose chases. Once, a digital hiccup took me 50 miles out of my way on a trip to an interview on the other side of Los Angeles. I lost my directional sense after just a few short months of depending on a digital knick-knack. Like most people who use their phones to ping their GPS coordinates off of mobile phone towers, I had outsourced my wayfaring abilities from my brain, and maybe the tiny magnets in my nose and eyes, to an electronic gizmo. Culture and technology overtook my biology.

A recent study of cab drivers in London offers a glimpse into how wayfaring corresponds to specific brain structures. In order to get their taxi permit London cab drivers have to memorize the convoluted rat's nest of English streets and be able to navigate the entire metropolis without maps. Researchers hooked up cabbies to MRI machines and discovered that the longer they drove cabs, the larger the volume of their hippocampi increased. The finding proved that thinking spatially can drastically change the brain. However, and perhaps more interesting, is that once the drivers retired, their brain structures returned to ordinary sizes.

It's strange to think about how quickly a human sense can disappear when it falls into disuse. Even when it is the most defining characteristic of a person, simply not exercising the neural muscle over a short time can make someone entirely dependent on external resources. While digital maps are a fairly recent phenomenon, there are other much more fundamental biological abilities that have simply vanished from our everyday use altogether. They lie there dormant and unused until the right stimulus wakes them up.

Some abilities stem from the deepest recesses of our animal past, from a time before our species walked upright. Perhaps before our amphibious ancestors crawled out of the water for good.

The story of the discovery of animal responses to water starts in 1894, when a physiologist named Charles Richet began tying up ducks so tightly with twine that they couldn't work their lungs anymore. The rather sadistic experiment could have been a plot point in a horror story, but it was actually an attempt to understand how water affects a duck's central nervous system. In a series of experiments that must have decimated the local duck population, Richet learned that the average duck can survive for about 7 minutes without breathing when on dry land. However, when Richet submerged the duck in cold water, the duck lived for almost 22 minutes before expiring. Somehow the sensory stimulus of being in the water had suppressed the doomed duck's metabolism. But it wouldn't be until 1962 that anyone tried a similar test on humans.

Luckily for the human subjects of Swedish-born physiologist Per Scholander, testing methods had gotten more sophisticated over the intervening century. Instead of binding his volunteers to death, he used lab equipment to measure the heart rates and oxygen levels of volunteers while they dove to the bottom of a pool. The second that the cool water covered their faces, Scholander noticed an immediate and corresponding decrease in their heart rates. It was the same phenomenon that Richet has seen with his ducks. In the second phase of the experiment Scholander installed a few pieces of fitness gear at the bottom of the pool and told his subjects to work out while underwater. While the toil of lifting weights underwater must have been awkward, the heart monitors showed that no matter how vigorous their activity was, their heart rates remained suppressed. Not only that, the volunteers also experienced vasoconstriction while beneath the surface of the water as the blood from their extremities shunted toward their core.

The water, it turned out, triggered a series of physical changes in his subjects that advantageously prepared them to survive in a hostile underwater environment where breathing would mean death. It's a reflex that Scholander named "the master switch of life," a term that has since been taken up by a burgeoning free-diving community that depends on these built-in changes in the body to extend the duration of their underwater forays. Freedivers, described masterfully in James Nestor's book *Deep,* are able to rapidly descend hundreds of feet below the ocean surface and then come up without risking the dangers of decompression that scuba divers have to worry about. As they dive, the pressure of the water collapses their lungs while the cold subdues their oxygen consumption. A world champion freediver can descend as much as 1,000 feet and still resurface alive and unharmed. In 2012 one freediver managed to hold his breath for a mind-boggling 22 minutes underwater during a competition.[1] This water-triggered master switch also has uses for landlubbers suffering from anxiety attacks or heart arrhythmia. If you happen to be prone to panic attacks, then submerge your face in ice water at the peak of the attack, which will signal your body to prepare for going underwa-

---

1   Readers beware: Some methods found in this book can help a person hold their breath for a very long time, but the techniques are not appropriate for use in deep water. The Wim Hof Method carries a small chance of unexpected fainting—which is safe enough on dry land, but not so much in the water. In addition, even proper freediving techniques can be dangerous. Between 2006 and 2011, 308 freedivers died in the course of training and competition.

ter and disrupt the heart palpitations.

It's not easy to know how many vestigial responses like these lay dormant deep within our human physiology. They are the sorts of abilities that only manifest themselves when the right conditions arise. They are the gift of millions of generations of incremental biological changes from ancestors whose daily challenges we only have a dim understanding of. Because most humans nowadays live within a narrow band of homeostasis, unlocking new responses occurs mostly by chance. And yet, when they do trigger, we are not always conscious enough to notice how they kicked in. In the modern world there are almost no people who are truly "untouched" by civilization. There is no control group that scientists can reference for what a wild human might actually be like. Instead the best we can do is rely on early anthropological literature, first-contact reports, and the mythologies and oral histories of native people around the world.

One story that I'm particularly fond of, even if it is also enigmatic, is the story of the last Apache warrior who stood up against the onslaught of Western civilization. After quashing the Confederacy in the Civil War, the Union Army turned its attention toward the Arizona territory, where a small band of Native Americans were ambushing settlers and small military units. By the mid-1860s most of the Apache had already been rounded up and placed into the camps that would eventually become the reservation system. But one particularly angry warrior named Geronimo resisted. He led a small band of fighters on a deadly campaign of bloody revenge for the atrocities committed against his people. Shutting down this renegade became a top priority for officials in Washington, and they committed 5,000 soldiers to the effort. They installed a revolutionary technology called the heliograph—a precursor to the telegraph that transmitted Morse code via flashes of light—to coordinate the campaign. And yet, despite what was effectively the largest manhunt in history, Geronimo's career extended through almost 30 years of violent conflict only to eventually surrender on his own terms when he was in old age. In that time he never had the advantage of numbers or technology, living instead by his wits and luck alone. Some historians record that his party killed as many as 5,000 people. The Apache credit Geronimo's success to divine favor. They believe the god Ussen granted him a special power that they called "enemy against," which let him know where his enemies were at all times. With it he could predict the weather and was said to be supernaturally in tune with the world around him.

While Geronimo's successes are a rare example of indigenous triumph over the juggernaut of Western progress, there are many tales of native healers and shamans who displayed similar abilities. The nomadic reindeer herders of Norway, the Sami, were said to be able to communicate telepathically with one another over impossible distances. Similar feats show up with certain Aboriginal groups in Australia and among various South American tribes. Whether any of these are innate human abilities, or are instead things that fall under the purview of faith and religion, is anyone's guess. Rigorous scientific studies of so-called psychic abilities have almost universally failed at proving the seemingly supernatural claims. That said, by the time modern science got around to even asking the right questions, the entire ways of life of native peoples had all but vanished.

The field of anthropology barely existed before the year 1900, and accounts of native populations before then often come from the perspective of the conquerors or missionaries—all of whom had vested interests in diminishing the accomplishments of the people they studied. The historical record only goes back so far, and attempting to gaze deeper into our evolutionary past, to a time before writing, is limited by what has been preserved by chance in the archaeological record. To get an idea of what we can know about how humans engaged with their environment in the deepest recesses of history, I call up Marc Kissel, a paleoanthropology postdoc at the University of Notre Dame. Marc also happened to be my roommate back when I enrolled in a PhD program in cultural anthropology in Wisconsin, and he has since become my go-to source for everything on human evolution. His research focuses on the origins of symbolic thought, the earliest evidence of violence between early humans, and the emergence and prominence of Neanderthals.

I ask him what challenges our ancestors faced and whether there was any evidence of physical abilities they might have had that we have since lost. He gives the scholarly equivalent of a shrug. "They were almost certainly more physically fit than us," he begins, "but the record isn't perfect. We don't even have a complete Neanderthal skeleton, just pieces from a lot of skeletons to make a sort of Franken-Neanderthal for us to understand their biology." The softer parts of their bodies, anything that might decompose over tens of thousands of years, has pretty much been lost.

Since we first started digging up their abnormally heavy bones in a limestone quarry in Belgium in 1829, Neanderthals have sparked heated debates about their place in human evolution. One thing is clear, however: Neanderthals are the closest species to humans that has ever existed on Earth.

Even so, they still technically sit on a different twig of our evolutionary tree. They appear in the fossil record as far back as 300,000 years and seem to have mysteriously gone extinct about 40,000 years ago. Some theorize that their decline happened in part because of the rise of our own species—either through warfare or just competition over resources. Either way, there was surely some contact between us and the Neanderthals, and it appears that the two species mixed and reproduced together. If you're of Asian or European descent you likely have between 1 and 4 percent of Neanderthal DNA in your own personal genome. Nonetheless, knowledge of early human and Neanderthal biology is pretty much limited to what we can understand from the bits of their bone fragments we've dug up. There is *one* thing that we do know, however: The invention of technology, as a rule, seems to correlate with a generally weakening of the raw physicality and resilience of our species.

"It was much colder then, and during the time that we didn't have fire they would have had to eat a lot of raw meat," Kissel says. "This almost certainly meant that they had some pretty impressive gut bacteria." If not for resilient guts humans would never have been able to survive on uncooked meat without risking serious illness. Indeed, the fossil record tells us that some of the most obvious skeletal changes occurred in conjunction with cooked food. Richard Wrangham, a biological anthropologist at Harvard, argues that the human jaw started shrinking once we learned how to control fire. Since cooking softens meat and kills potentially harmful bacteria, we no longer needed the pronounced mouths and powerful forward-jutting jaws of our more apelike ancestors. Cooking also made vegetables and meat more nutritious. Instead of spending most of our time chewing plant fibers to break down resilient cellulose we could outsource that effort to fire, which radically increased our ability to extract calories. Wrangham writes in his book *Catching Fire* that the invention of cooking by our early ancestors was literally what made us human and sparked the evolution of *Homo erectus* 1.8 million years ago.

Homo erectus was different than the great apes that preceded it. *Homo erectus* walked upright on two feet, and, because digestion takes less time when food is cooked, has a smaller gut than most apes. Over the next million years or so *Homo erectus* experienced a gentle progression into the more refined human form we now enjoy. Presumably because *Homo erectus* didn't have to spend so much time chewing, it could do other things that led to evolutionary pressure that increased reliance on technical skill; therefore, brain size increased. At some point, *Homo erectus* also lost its body hair, seemingly because he could use fire to keep himself warm at night.

Wrangham's analysis breaks new scientific ground precisely because he shows that technology has had a profound effect on the human form. In fact, its effect is so strong that it is almost impossible to separate human evolution from cultural and technological changes. To put it another way, the body you have now would not have been possible had we not invented fire. But that doesn't mean the transitions were always smooth, or, for that matter, fast. At various points, as technology and biology altered course together, when the pace of change was too rapid, we began to suffer from evolutionary mismatches between our outdated biology and the world we were quickly inventing.

The most classic example of an evolutionary mismatch—and an easy one to locate in the fossil record—is the position of our teeth. As the human mouth shrank over time, the number of teeth remained constant. Today, as we grow older, wisdom teeth poke out of our gums and shunt the other teeth out of position, which requires surgery or intensive orthodontics to correct. As such, humans are the victims of their own success. Before cooking, humans spent a lot their day chewing on food to break it down, and the process moved teeth around the jaw and put them into their correct positions. But since fire made food softer, human jaws didn't have to chew as much and, absent the constant pressure, teeth could shift out of position. Marc Kissel says he knows of no other animal, certainly none of our ancestors, that has the same problem. Indeed, cooking also seems to have changed the quality of our teeth, not just their position. As an archeologist Kissel has examined hundreds of hominid teeth from both hunter-gatherer and agricultural societies. The latter have more cavities. The mouths we have now are in much worse repair than those of early humans. Dietary changes, most menacingly sugars, make us good candidates for tooth decay. No Neanderthal or *Homo erectus* ever had cause to visit a dentist.

Understanding changes in soft tissue that give rise to evolutionary mismatches is a far less straightforward affair. Barring an amazing stroke of luck, like discovering a flash-frozen Neanderthal body in permafrost, there is no definitive way to tell how our muscles, brains, fats, and organs have changed over time. Anthropologists and biologists are stuck making educated guesses about the vast majority of our early biology.

Neanderthals thrived across Europe during an ice age in which glaciers extended much farther south from where they reach now. Located far from our arboreal ape ancestors, Neanderthals created a sophisticated culture, buried their dead, and used ornamental beads to decorate their clothing and jewelry. Though less durable artifacts haven't survived in the record, most anthropologists believe that at the height of their technological proficiency Neanderthals would have used heavy furs and built semipermanent dwellings that they could heat with fire. Those technological advantages, however, don't explain how the species persisted for more than 200,000 years in such a cold environment.

In 2002, Theodore Steegman, a professor of anthropology at Tulane University in New Orleans, was contemplating his forthcoming retirement when he penned an article in the *Journal of Human Biology* that proposed an entirely new way to think about Neanderthal survival. Steegman reasoned that Neanderthals must have relied on a variety of biological strategies to survive in the cold, and he set about to examine the scientific literature to find even the most obscure anatomical research on modern humans. He was looking for clues in human physiology that might help him determine what the meat of a Neanderthal body might have been made of. By measuring the skeletal structure of various specimens he estimated the approximate muscle mass of the Neanderthal species. Steegman noted that their shorter limbs and greater general body mass would have naturally insulated their core from harsh weather. This observation was hardly revolutionary and would have gone unnoticed if he hadn't also suggested that Neanderthals used brown adipose tissue (BAT), colloquially known as "brown fat," to keep themselves warm.

Brown fat is a fundamental tissue in mammals that, at the time, was mostly understood as something that rodents used to help heat themselves during hibernation. The spongy, fatty tissue looks a lot like ordinary white fat that most mammals use to store excess caloric energy. But where white fat can serve as an insulator, brown fat has an active role to burn white fat to generate body heat. It's the only mammalian tissue whose sole purpose is to make heat, or, in scientific terms, thermogenesis. In humans, however, BAT was only considered important in newborns. The very first challenge that a human faces once it is born is the constant fight to keep a stable body temperature. With a relatively high ratio of surface area to body mass, babies lose heat much faster than adults. This is why many premature babies spend their first weeks in incubators. The easiest way for an adult to rapidly increase their core temperature is through shivering. The activity of muscles generates a moderate amount of heat as a byproduct of movement. Fresh from the womb, infants lack significant musculature and can't shiver themselves warm. Instead they are usually born with a layer of chubby rolls of insulating white fat. When their core temperature begins to drop, BAT turns on, sucks white fat from their system, and releases a cascade of heat energy.

Once the infant becomes a child and its muscles begin to tone up, the white baby fat disappears and along with it the brown adipose tissue. By the time they're adults, most humans have very little BAT in their system. What does remain is often just a few teaspoons' worth of the tissue along their spine and shoulders. Most doctors figured that BAT was simply irrelevant for adults. In fact, most adults have so little BAT that anatomists didn't even know it existed in us until the 1970s. However, Steegman argued that Neanderthals were able to deploy the same heating strategies as human infants, and he scanned obscure literature on human physiology to prove it.

He cited a Finnish study from 1981 in which an anatomist compared autopsies of two groups of manual laborers. The workers came from two different groups: people who worked indoors throughout the year and others who toiled outdoors. The Finnish winters are famously brutal and closely mimic the conditions that Neanderthals would have faced throughout the winter. The study compared BAT concentrations in the two sets of laborers and found that the ones who spent most of their time out of doors had significantly higher deposits of it. They had so much of it, in fact, that proportionally it resembled the levels found in wild mammals. The sole exception was a single worker who died during the summer. Steegman extrapolated from this that BAT levels vary with the seasons, and that if we regularly exposed ourselves to the environment humans would build up deposits of BAT in cold weather and then shed it when it gets warmer. Steegman realized that humans in industrial countries generally lacked the conditions that would spark the body to start building BAT. It's not that the levels of BAT naturally decline as humans get older, it's just that the body learns that it doesn't need to invest energy into building BAT when it lives in perpetual summerlike conditions indoors. Despite some of their technological achievements, Neanderthals were still subject to the seasons, and it would have made a lot of sense (if they were anything like humans) that the near constant exposure to the cold would have built up their levels of BAT.

Like a lot of groundbreaking science, Steegman's study was too far ahead of its time, and so it went mostly unnoticed by the academic community. This was at least in part because measuring BAT levels in live humans is an incredibly difficult task—one that wasn't even really possible when Steegman was writing.

Even so, Steegman wasn't the first person to try to decipher the secrets of body heat. Anthropologists started conducting extensive surveys on cold adaptation on indigenous people around the world beginning in the 1930s. Back then the procedures were often rudimentary and often involved a sunburned researcher in a pith helmet convincing some unsuspecting tribesman to wear a rectal thermometer while standing out in the cold or jumping into ice water. One can only imagine the pantomime involved in such a transaction. However, it was all in the name of science and whatever the interpersonal interactions, researchers found a surprising array of biological strategies that native people deployed to stay warm.

Above the Arctic Circle in the Norwegian territory of Lapland, one scientist found that the local reindeer herders had an incredibly strong vasoconstriction response to the cold while at the same time exhibited a pronounced drop in their core temperatures when they slept out in the snow at night. The Lapps, it seems, had discovered a way to simply not be bothered too much by getting cold, and had a much higher threshold before succumbing to hypothermia. But the strategy for surviving the Arctic climate was found not to be the same across all polar-dwelling populations. Inuit subjects who lived in northern Canada showed significant heat loss when given the same test, leading researchers to suspect that they were instead able to passively increase their metabolism. Inuit were also able to keep their hands warm when immersed in cold water, which means they had a blunted vasoconstriction response. This adaptation strategy would have meant that their bodies intentionally sacrificed some of their core temperature in order to have better dexterity with their hands.

An anthropologist named H. T. Hammel who studied Aborigines of the central Australian desert in the 1950s observed that even though the tribesmen wore no clothing, they still slept outside throughout the winter when the temperature dropped to just below freezing at night. They slept on bare ground with only a screen of bushes to protect them from the howling winds. While they were asleep, Hammel tested the Aborigines' skin temperature and found them cold to the touch compared with the Western researchers who were there studying them. This meant that the Aborigines not only had intense vasoconstriction, but their bodies somehow (and mysteriously) also experienced less heat loss than the European control group. Meanwhile, in Africa, Kalahari Bushmen in strikingly similar conditions neither shivered nor showed any signs of vasoconstriction, meaning they kept a normal skin temperature all night long.

Studies of the Inuit, Lapps, Kalahara, and Aboriginals reveal that the human body isn't limited to a single strategy to deal with the environment but can pull from an array of solutions to extreme weather.

While certain anthropologists would have been overjoyed to spend the entirety of their careers prodding and poking indigenous peoples around the world for the treasure trove of data they represented, the golden era of biological measurement dried up by the 1960s. Whether by conquest or the simple lure of Western convenience, by then there were already too few untouched populations in the world to gain reliable data on what a wild human biology might look like. Lacking PET/CT scans, it was also impossible to determine the exact causes of increased metabolic activity in native people when they found it.

In broad strokes, however, thermogenic research has showed four main strategies that the body can use to resist the cold: Humans can increase their metabolic rate by shivering; they can stay warm through some passive metabolic mechanism; or they can shunt blood to their core by closing off the arteries in their extremities. Lastly, as in most cases of people who live in colder climates, the human body can also accumulate higher levels of ordinary white fat as an insulation to help stem heat loss in the first place. Anthropologists credited the different strategies to genetic adaptations that evolved from people reproducing in diverse environments and not necessarily a sign of evolutionary abilities shared among all humans. However, by 2009 Steegman's hypothesis about brown fat was about to find an unlikely champion when new science had the potential to turn the entire field of human thermogenesis on its head.

By then PET/CT scans were becoming a regular addition to cancer centers around the United States and Europe. However, oncologists were beset by an unforeseen technical problem. The scans are a sort of nuclear imaging test that are sensitive to metabolic activity. To take one, doctors inject their patients with a radioactive dye that circulates throughout the body and gets taken up by particularly active cells—something that can be useful to identify early stages of cancer. In almost real time, the scanner maps out where the dye ends up. However, when looking for cancer indications, about 7 percent of PET scans returned blobby images of cancerous-looking areas around the shoulders and thorax. But when the doctors performed biopsies on the danger spots, they never found what they were looking for. So a Harvard researcher named Aaron Cypess looked at 3,640 different scans to figure out what the blobs really were. After comparing the images with obscure textbooks on human anatomy, he discovered that the blobs mapped perfectly over the areas that an anatomist back in 1972 identified as "inactive" brown fat deposits.

Cypess deduced that the very examination rooms where the PET/CT scans happened were ideal for activating BAT. They were abnormally cold, and patients wore only thin hospital gowns when they were in the confines of the machine. Under these conditions BAT was simply doing its job to keep people warm: sucking up fats and sugars from the bloodstream and producing enough heat to light up the PET/CT readouts.

More important, the discovery showed that BAT wasn't a vestigial tissue at all, but that even adult humans might reap benefits from its presence. And thus, BAT might not only help explain the mysteries of how Neanderthals survived through ice age winters, but also how modern humans could use it to get through their own winters as well. Oncologists now note that PET/CT scans taken in the winter months have more instances of BAT than in the summer, suggesting that it comes and goes with the seasons.

At this point, it's worthwhile to mention another story about first contact between Westerners and an indigenous group. In December 1620, a group of religious outcasts from Europe landed on a foreboding spit of land on the coast of what is now called Massachusetts with the intention of setting up a colony. The settlers arrived in three rickety sailing ships and disembarked in the middle of a harsh New England winter. The colonists built a crude settlement and shivered unhappily in their drafty huts, unsure if they would be able to make it through to spring. They were alone for months until a tall, long-haired Indian walked into their camp on a windy March day wearing only a fringed loin cloth. He greeted them in English with the words "Welcome! Welcome! Englishmen!"

While the story of how Samoset knew any English at all before any Europeans had settled in North America is interesting (spoiler alert: he learned it from fishermen who had come to North America to hunt for cod), what is perhaps even more mind boggling and relevant to the topic at hand was his outfit. Being nearly naked in the Cape Cod winter is, to most people's minds, crazy. And the first thing the Pilgrims did was offer him a coat. He took it with a shrug—who wouldn't want to be warmer in the winter?

However, after Samoset introduced the Pilgrims to the local Wampanoag tribe, it turned out that the same basic garb was standard throughout the entire region. Their secret didn't rely solely on their genetics. Contemporary accounts by colonists stated that the local tribes relied on conditioning to make their children more robust. Every winter they placed their infants and children in the snow for a few minutes every day before bringing them back inside their houses. Repeated exposure helped condition the children to become comfortable in environments that would make the typical colonist want to die. Today, the state seal of Massachusetts depicts an Algonquin native in a bare bones winter outfit, an image that no doubt stuck in the minds of the Pilgrims as they considered how strange the land was in which they had chosen to live.

The Wampanoag seemed to understand intuitively that resistance to the elements wasn't an innate power. Instead, they recognized that their bodies already had all the tools necessary to survive in the environment if they made a conscious effort to adapt.

It's the same fundamental biology that probably allowed Neanderthals to survive, and there's no reason to think that these abilities are not shared by every person alive today.

It's a lesson that hasn't been lost on Ray Cronise, a former NASA scientist living in suburban Alabama. He spent 15 years supervising experiments at the Marshall Space Flight Center, but his career took a turn once he decided to develop a scientific way to shed weight that wasn't focused around counting calories. Standing 5 feet 9 inches tall and weighing 209 pounds, he wanted to trim down to a fighting weight of 180, but in the process of investigating his own metabolism he ended up developing a theory that cut harshly against the grain of how most of us live in the modern world.

Every human alive today lives in a cocoon of consistency: an eternal summer. "We're overlit, overfed, and overstimulated, and in terms of how long we've been on Earth, that's all new," he says to me while summer is settling in Alabama. We're missing out on what he calls "metabolic winter," a time when the body adjusts to discomfort and scarcity between times of plenty. As he writes in his opus article on the subject, "Our 7-million-year evolutionary path was dominated by two seasonal challenges—calorie scarcity and mild cold stress. In the last 0.9 inches of our evolutionary path we solved both." The inevitable result of losing seasonal variation is obesity and chronic disease. As proof, he doesn't point only to the population of his home state, which ranks at #5 of the most obese states in the nation, but also to the fact that our pets are fat, too. "The only two animals in the world that suffer chronic obesity are humans and the pets we keep at home," he says. "There's a connection."

The key to fixing the problem, according to Cronise, is to artificially reintroduce seasons back into our lives. Drawing on the growing literature on brown fat and metabolic analysis, he prescribed himself daily hour-long walks in sub-60 degree temperatures along with moderate exercise and calorie restriction. In the course of just 6 months, from June to November, he dropped almost 40 pounds.

Cronise's research isn't just about losing weight, though. He's trying to correct what biological anthropologists call evolutionary mismatch diseases, or, in layman's terms, what happens to the body when the pace of technology overshadows our fundamental biology. By looking backward at where we came from and the conditions that our ancestors inhabited, we might be able to create more adaptive conditions to flourish ourselves. What's more, because humans are inherently adaptable, even moderate changes in the way we approach the environment should trigger our own evolutionary programming. Just about anyone should experience rapid results. And whether that means simply becoming more robust in life in general or unlocking untapped mechanisms in my own body, it is a challenge that I figure I can take on myself.

3

MEASURING THE
IMPOSSIBLE

ROB PICKELS HAS small white patches in his mouse-brown hair and
the athletically trim body of a guy who spends 16 hours a week biking
up and down the mountainous roads around Boulder, Colorado. As the
lead exercise physiologist at the Boulder Center for Sports Medicine, he
helps train and analyze world class athletes as well as patients who are
fresh out of the cardiac unit at the nearby hospital. A few weeks before
our meeting I'd sent him a long and scattered e-mail about the feats that
Wim Hof seemed to be able to accomplish. I regaled him with the story
of my dramatic weight loss in Poland and how I ended up on the top of
a mountain in little more than my skivvies. I ended the message with a
small request: Would he be willing to assess my own physiology as I
undertook Hof's method for 6 months?

The subject is intrinsically interesting to Pickels, as he is always looking for new ways to give athletes an edge in competition. His own history
with hacking the body goes back to a time when he attended Ithaca College in upstate New York as one of the school's top 400-meter hurdlers.

At the time, he was studying human physiology with his eye on eventually taking graduate courses in sports medicine. One day his professor mentioned a metabolic process called the bicarbonate buffer system that regulates the amount of carbon dioxide and hydrogen in the blood stream. This complex system dictates that when the body works harder, acidity builds in the muscles. We experience this as fatigue. Pickels figured that if he could tweak the equation between exertion and the bicarbonate that buffers acidity, he could improve his performance in the races he was competing in.

Loading his body with baking soda (i.e., sodium bicarbonate) was an easy way to do this. The only problem, as he saw it, was that he'd need a lot of baking soda to make it work, almost 5 heaping teaspoons' worth. That is enough to give someone a long, unpredictable, and explosive case of diarrhea, which he considered an acceptable trade-off. When we finally meet in his office, I ask him if he thought his experiments with baking soda were worthwhile. "It might have just shaved five-tenths of a second off my race time, but in an event that lasts less than a minute that is the difference between first and last place," he replies. Pickels never lost a race in his entire college career.

That is the sort of experimental attitude I want to see in anyone who plans to run me on a treadmill until I almost pass out. We start our testing sessions in the summer of 2015, 2 and a half years after I initially met Wim Hof. During the intervening years I'd moved to Colorado from California and spent most of that time in different desk chairs researching and writing a book on the sometimes-fatal side effects of intensive meditation.

Now I am ready to investigate the other side of the meditation equation. I plan to dedicate myself to at least 6 months of Hof's training regimen in the hope that I might see some stark changes in my own biology. Before I start in earnest, I need to find a way to track how my body changes over time. It is May, and the thunderstorms rolling up and down the Front Range of the Rocky Mountains aren't bringing any more snow. In a way, it is the perfect season to establish a baseline. Pickels agrees to design a test before I ramp up my training in the Wim Hof Method to see how—or, rather *if*—my body responds to a steady regimen of heavy breathing, meditation, and cold exposure. Pickels has seen crazy body hacks before, but I can tell that he is skeptical of the program that I am suggesting.

Fortunately for me, Pickels isn't interested in my core temperature—which means I get to forego the rectal thermometer he no doubt has stashed in a nearby supply closet. Instead, he wants to know how my body stores and uses energy. We're going to measure my metabolic rate by hooking me up to a plastic snorkel and pushing me hard on a treadmill until I can run no longer. There's a part of me that wonders if perhaps he'll find something special, and that maybe despite my infrequent exercise and desk jockeying nature he might find that I have a hidden metabolic talent. When I drop the idea by him, he smiles and launches into a story about Evelyn Stevens, a Wall Street banker who worked 50-hour weeks with little more than an occasional jog to stay in shape. Stevens borrowed a friend's bike on a lark, donned some Spandex, and toured with a local bike club one weekend. She was more surprised than they were when she left her fellow riders in the dust. A few years later she won two stages of La Course, the women's Tour de France, and currently holds the women's record for longest distance traveled in a single hour on a bike.

As Pickels inches up the speed on the treadmill I ask him if maybe I might hide the same sorts of talent as Stevens. He smiles like he gets that question a lot.

"No," he says. "It's mostly about genetics." From the moment I walked into the offices in Boulder, he knew that I would never be world class. The dead giveaway, he says, is the one muscle in my body that I always thought I had legitimate reason to be proud of. After a splinter lodged into my heel as a kid I've had a tendency to walk on my toes. My loping and oddly bouncing stride caused no shortage of middle and high school teasing, but as a side effect it also gave me abnormally large calf muscles. While they are by far the most toned part of my body, Pickels tells me they're also a clear predictor that I just don't have what it takes to be an endurance athlete. "It's like you're carrying two pendulums at the bottom of your legs," he says. "You have to work harder than someone born with long, skinny legs."

According to the guy who sees hundreds of top sportsmen in his office every year, world-class athletes are born, not made. He then tries to ameliorate my dashed dreams by saying, "Well, look at it this way. If you won the lottery and wanted to train with us every day, we could probably make you relevant in an age-appropriate tier. Maybe not a professional tier, but definitely higher performance." So there it is. Apparently I've got at least a little sliver of hope of being an Adonis after all. Consider me challenged.

Pickels increases the speed on the treadmill a few notches and I breathe heavily through the snorkel that connects to a pricey piece of hardware that measures the mix of carbon dioxide and oxygen coming out of my lungs. In the course of an hour the treadmill test pushes me harder and harder, and every 4 minutes Pickels pricks my finger with a lancet to draw blood to analyze for lactate content. Eight pricks later and my finger looks like it just lost a fight in a local bar. It's bruised and bleeding, and I'm pressing on it with a swab of cotton to force a clot. I feel like I can taste blood in my mouth.

Pickels has me rate my exertion at different stages of the process. I flash him moderate fives and sixes out of ten when he has me running up a steep hill at a 9:40 mile pace. The data that he's reading, however, tells a different story; it says that I'm getting close to collapsing. My heart rate hits 182 and I'm breathing hard. He says that he wants me to reach a pace where I don't feel like I can run any faster, and presses a button on the computer that sends the treadmill on an even steeper incline. It feels like he's planning to send me up one of the Flatirons—the jagged mountain formation on Boulder's flank that demarcates the eastern edge of the Rocky Mountains. Three minutes later I'm tapped out. I groan through the snorkel and make a hollow echoing sound of a drowning Muppet. That's it, I'm done after six stages of incrementally harder exertion. My vision is blurry and I want to put my head down on the rail of the treadmill to recover.

Pickels examines several thousand data points that track my metabolic rate. Information splashes across two computer screens. There's predictable curve upward that matches my exertion with every increase in speed. From the gasses that left my lungs he can tell that I burn mostly carbohydrates, indicating that I'm definitely not a world-class endurance runner. Top athletes that compete in ultramarathons run mostly on slow burning fat for a reliable source of energy. I'm the exact opposite; you could call me a carb monster, someone who seems to have granola bars in his bloodstream. It's the sort of cheap and easy metabolic activity indicative of a rather ordinary American diet. He almost chuckles when he tells me that I still have room for improvement.

That said, it's not like I'm in *bad* shape. Since first meeting Wim Hof, I now work out two or three times a week, usually a 2- or 3-mile run around a lake. I might do an hour of lap swimming or take an occasional hike in the nearby mountains. I weigh about 190 pounds and have no serious underlying medical conditions other than occasional autoimmune mouth ulcers that I've had since I was a child. In other words, I'm pretty ordinary.

In order for the experiment to work I need to keep my routine roughly the same for the next 6 months. I'll keep up the same semi-regular exercise pattern with two or three workouts a week, and I'll eat the same sorts of foods. The only change I plan to make is to do the Wim Hof Method as regularly as possible. Other than a few excursions I have planned, everything else should be the same.

So it's going to be an odd year. On one hand, I need to maintain a certain level of fitness so that I can actually measure the effects, if any, of the Wim Hof Method. On the other hand, I'm going to explore different parts of the world where people use the environment to train their body. I want to slog in the mud with the Spartans and train with people crazy enough to surf monster waves. I even have plans for trying to persuade Hof to take me on one of his death-defying expeditions.

The workout regime itself is relatively simple. Every morning before breakfast I'm going to do breathing exercises, followed by breath-hold pushups and a headstand. It's a truncated version of the method that Hof recommends for a full hour a day. I choose this abridged routine because it is more manageable and easier to keep track of. When I do work out, it will be outside. For now that means shirtless runs in the sweltering summer heat, but as the seasons gradually progress to fall and winter there will be snow, wind, and hopefully a liberal buffeting of frost. Finally, I'll take cold showers every morning and, when possible, lie down in the snow like I'd learned to do in Poland. All together, the routine will add between 15 to 20 minutes of exercise to my day and should be easy enough to keep up with throughout the 6 months of the experiment. The key, I think, will be to keep exposing myself to temperature variation and getting comfortable with whatever weather is outside my window. I want to be in touch with the seasons. My goal isn't to become the next champion ultramarathoner or to build muscle mass or even simply to lose weight. I want to see the effects of introducing variation into my environment and discover what happens when I try to take control of my body's autonomic processes.

"From my perspective I'm mostly interested in your metabolism," Pickels tells me, noting that cold immersion might well change the way that my body utilizes fat. Then again, noting the pendulums that I have for legs and the general carb-hungry tilt to my blood, he's not exactly holding his breath for a miracle. He asks me to report back in 6 months and says that he's looking forward to analyzing the results.

Whatever the results, they'll offer a peek into a small, measurable slice of what the Wim Hof Method might be able to accomplish. But there's another goal that will be harder to pin down even with a million data points: finding the critical fault line that helps crack open the space between the conscious mind and my unconscious physiology. If it exists, and if I'm able to tap into it, I expect I'll only be able to explain it once I feel it. I call it the wedge.

4

THE WEDGE

THE FRONT RANGE of Colorado is a string of mountains that marks the boundary of the Great Plains and the Rocky Mountains. East of here the land slopes ever downward through prairie lands and bayou until it reaches the Mississippi River. To the west, spires of solid rock burst forth from the ground like a fortress, the result of millions of years of tectonic action. In 2015 the Front Range endured the most brutal winter that anyone could remember. Last summer, I'd rented a small ranch house in the city of Boulder. All was well. Then came winter. In a single morning snow could pile up past my knees, making it hard to move more than a few feet outside the door without first reaching for a shovel. For some it was a bleak time of year. My small gray and black tabby cat abandoned hope of ever going outside again. In due time, though, spring settled upon the eastern slope and with it green grasses, thorny weeds, and wildflowers awoke from their dormancy. I take in a breath of that fresh spring air and suddenly realize that a certain species of birch tree that I happen to be allergic to has also emerged from its winter slumber.

The invading pollen of this tree saturates the air and wages war on my sinuses. Unable to stem the flow of allergens, my itchy eyes begin to water and I start to feel the irresistible urge to sneeze. So I let it go. A second later I sneeze again. And then again, until the explosions string together like an absurd form of Morse code. Sneezing is one of the body's many defensive responses against environmental threats. The neural circuitry between the nasal passage and the sneezing center in the brainstem is only a few inches long, and it is one of our most primal reflexive pathways. While sneezing can be pleasurable in small doses—indeed the brain patterns of a sneeze recorded in an MRI resemble mini-orgasms—an unremitting cascade of them is no fun.

It doesn't take long for the sequence of ceaseless nasal explosions to turn irritating, so I decide to make it stop. And just like that, it does. Everyone has the ability to resist a sneeze, and while it might be hard to describe this with words, we all know how to do it. There's no exact method except to say that you tighten something in your mind, or maybe think non-sneeze-worthy thoughts. Somehow you just will it to stop. In this moment of mental acrobatics you short-circuit an autonomic reflex and extend a measure of control over your body's programming. Of course, it's not foolproof—my allergies won't dissipate until the birch stops breeding—but recognizing this mental trick is the basic building block of taking control over your autonomic nervous system.

Call it what you want: willpower, focus, or concentration. The mental state you go into while trying to delay a sneeze is a sort of wedge between the autonomic and somatic nervous systems at the point where an environmental stimulus meets an innate response.

It's the same wedge that a person uses to calm their nerves when standing in the snow, withstand shivering, hold their breath just a little bit longer, delay an orgasm, stop feeling ticklish, or hold back the flow of urine as they search for a bathroom. It seems like a small thing, but it's a window into the root of a human power, and a place that, if exercised, can help unlock the body's hidden biology. Freedivers who descend hundreds of feet below the surface of the ocean on a single breath sometimes call it the "master switch": It's the point where the body meets the mind.

Any preprogrammed physical response is potentially susceptible to the wedge as long as it has three key characteristics. First there needs to be a clearly identifiable external stimulus. Second, that stimulus must trigger a predictable automatic biological response or reflex. Third, that physical response must elicit a feeling or sensation you can visualize or imagine independently of the external trigger. If the reflex has these characteristics, then using the wedge is as simple as setting up an environmental stimulus and then resisting the sensation that it triggers. Over time it becomes easier to maintain the tension between reflex and mental control.

That said, not every reflex is an ideal candidate for training. Allergic responses might make for interesting test cases, but like many autonomic functions, they exist for a reason. Sneezing removes allergens from the body and is part of a preset program that has enabled species to survive until today. Suppressing allergic reflexes could lead to all sorts of physical complications. The same goes for learning to hold your bladder at bay indefinitely. Sure it's possible, but it is not generally a good idea.

So training starts with one of the most fundamental human reflexes: the urge to breathe. When the Buddha first taught meditation to his followers, he recommended that they start by watching their breath move in and out of their body. Breathwork is a staple of every yoga class, as students move their bodies in sync with their lungs. The Wim Hof Method tasks students to hold their breath until they can't take it anymore. And then hold it just a little longer. This is the quickest and probably safest way to build your own wedge.

The urge to gasp for air is not directly linked to the amount of oxygen in the bloodstream. That's because, for some reason that has been lost in the convoluted process of evolution, the body cannot sense oxygen, only its byproduct. Breathing is a two-part process—inhaling to bring oxygen to the lungs and exhaling to expel carbon dioxide (CO_2). When the brain senses too much CO_2 in the bloodstream, the chest tightens, vision blurs, and just about every muscle from the abdomen to the forehead clenches down hard. When we talk about this sensation we usually say that we need to take a breath. However, on a physiological level your body wants to expel CO_2. It sounds counterintuitive but it's easy enough to test. Take a deep breath in and hold it until you feel the urge to breathe. Then release a little bit of air. With less CO_2 in your lungs you will feel like you can hold your breath a little bit longer. That's because you've removed a potentially poisonous waste product from your body and your nervous system has turned off the alarm bells.

This basic gas exchange creates an opportunity to trick your nervous system into extending the amount of time that you can hold your breath, thus leading to the very first training technique to crack into your nervous system. But before you try any of the techniques in the coming pages I need to offer you an obligatory disclaimer: The methods in this book carry some chance of injury or even death. When you are choosing where to exercise, bear in mind the risk of briefly passing out and falling down. Before trying any of these exercises, you should get your doctor's approval.

GETTING STARTED WITH BASIC BREATH-HOLDING

First you will need to establish a baseline. Take a deep breath of air and time yourself with a stopwatch to see how long you can hold it. Most people can hold their breath for between 30 seconds and a minute without any training, but everyone's underlying physiology is a little different. While you're at it also figure out how many pushups you can do. When I first started, my arms began to buckle at around 20. Some people can only do one or two. Whatever it is, write down your baseline, perhaps on the inside back cover of this book, so that you can check back on it later. I should note that this is a variation on the Wim Hof Method. It is an even more basic starting point than where he usually begins students with their training, and is not precisely the one that he teaches in his courses.

The first goal is to prepare your body by blowing CO_2 out of your system. Start by sitting down on a couch or lying on the floor and taking 30 fast breaths. Each inhale should take about a second, but you should not force the exhale. Instead let it flow out naturally. Your breath should sound like you're at the peak of a sprint just a few seconds shy of having to slow down. Pretty soon you're going to feel a little dizzy; this is normal. You might also feel tingling in your hands and feet. You might also feel cold. Or experience ringing in your ears. After about 30 breaths you will have radically increased your blood oxygen saturation and cleared out most of the CO_2 in your system. Now, finish the hyperventilating[1] with a big gulp of air and hold it with a full chest. Time yourself to see how long it takes you to reach the point where you feel the need to gasp. Hold on as long as you can and clench the muscles in your chest, arms, and legs. You're probably making a pretty crazy face as you struggle against the urge to breathe. But when you can't stand it anymore, slowly let the air out of your lungs. This gives you a little boost of time before you take any more air in. The first time you try this probably won't produce any dramatic results, but you should find that you can hold your breath a little bit longer than your baseline. You might also be surprised by how much physical effort breathing can be. The muscles around your diaphragm don't get a lot of strenuous, focused exercise and it can take a little while to build them up. However, most people see a dramatic improvement in how long they can hold their breath when they repeat the cycle of hyperventilation and breath-holding for three or four cycles—or for a total of about 10 to 12 minutes. And now comes the really cool part.

---

1   Throughout this book I use the word "hyperventilating"—which literally means fast breathing—in the context of Wim Hof's breathwork to describe his version of deep, controlled fast breathing. It should not be confused with the shallow, uncontrolled rapid breathing that is the signature of a panic attack.

POWER PUSHUPS

Start while lying with your back on the floor, then do one more cycle of about 40 breaths. Make the last 10 even a little faster. Then take a giant breath in, turn over, and immediately start doing pushups while holding your breath. Try not to think about anything else other than the count. Or, even better, think of nothing at all and have someone else count for you. You will feel that these were the easiest pushups you've ever done, and most likely will be able to smash through the pushup baseline that you set earlier. And you'll have done it without actually breathing. This will work with just about any sort of strength exercise—pullups, situps, weights, leg lifts, or dips—if for some reason pushups aren't your thing.

Of course, this is a pretty well-understood body hack. When you clear your lungs of CO_2 and fill them with air, there's enough stored oxygen to do a fair amount of physical exertion. However, your brain doesn't intrinsically know how you've changed the oxygen baseline to perform a little bit better. Every time you do this you effectively grow a slightly stronger wedge.

Another—more powerful—way to do these exercises is to start doing pushups after an exhalation with empty lungs. When you feel the urge, take a breath and continue until you are exhausted. These two methods work on slightly different parts of the nervous system.

Both of these exercises work to find the meeting point between the body and the mind. Your brain has a subconscious map of what it thinks are its personal limits based on previous experiences. That map isn't particularly well-drawn for anyone who isn't already the sort of athlete who constantly pushes themselves to the edge of their physical abilities. Since most people don't spend their days hyperventilating and holding their breath, their nervous system hasn't yet mapped out what is supposed to happen when they start. As it learns its new physical capabilities, new neural connections form and the brain creates a new set of guidelines. Congratulations: You are now training your brain to cede a little control to your conscious mind.

There are a few known dangers to this sort of breathwork. It's perfectly normal to feel a little dizzy, or for your hands and feet to get cold or tingle—both are symptoms of increased circulation. The body will literally get high on excess oxygen, and the constant air circulation in the lungs tends to cool the blood down as well. It's much less common, but possible, to faint while doing controlled hyperventilation or during a breath-hold. It is slightly more common to pass out during breath retentions when your lungs are full—while empty lung retentions will prompt the gasp reflex earlier. If you fall unconscious, your body will quickly reset to its automatic programming and you'll wake up within a minute. In my experience there is an increased chance of fainting while doing the exercises with full lungs, since the increased volume of air throws off the way my body detects CO_2. Once, after reaching 80 pushups, my arm collapsed and I banged my forehead on the wood floor. The bump on my head was an indication that I'd found my limit.

Some medical literature has also shown that it's possible to have a stroke or even a heart attack from hyperventilation so, again, people with serious heart conditions or advanced circulatory problems should take extra care. And while it would seem that this sort of breath practice would help people hold their breath underwater—indeed some freedivers employ remarkably similar techniques—there is always the risk of pushing your limits too far and drowning. As previously noted, 308 freedivers drowned while training and in competition between 2006 and 2011. There have been no fatalities among those practicing Wim Hof's breathing methods on dry land as far as I am aware.

The autonomic nervous system divides itself into two interrelated components. First is the sympathetic nervous system, which controls the so-called fight-or-flight response. If you were a car, the sympathetic system would be the gas pedal. It gives you short-term boosts in energy, activates the adrenal gland, and triggers dilation and vasoconstriction. The parasympathetic nervous system controls the opposite responses, sometimes called the "feed-and-breed" actions. This is the equivalent of a car's brake pedal. The parasympathetic nerves act on digestion, saliva-tion, sexual arousal, and tears. Both systems are involved in breathing, and the right techniques can help strengthen both sides of the nervous system. In the same way that someone at a gym can focus on particular muscle groups during exercise, different breathing routines target differ-ent aspects of the nervous system. The basic breathing technique of hold-ing your breath after hyperventilating acts mostly on parasympathetic nerves. The sympathetic system only starts to kick in closer to the moment when you are struggling to stop yourself from gasping, and it is often best to train with breath-out holds. Breath-in holds are better for getting to your absolute maximum number of pushups or retention dura-tion, but are less efficient at cracking into your nervous system.

POWER BREATHING

In order to prime your body to get to the gasp point earlier (and thus build a stronger wedge and activate the sympathetic nervous system), start with the basic breathing method for approximately 30 quick, deep breaths. Keep your eyes closed and breathe hard enough that you begin to feel light-headed. Now, instead of taking in a deep breath and holding it, let most of the air out of your lungs like you would at the end of a normal breath (by which I mean, don't force it) and hold your breath with mostly empty lungs. Your body will quickly deplete the oxygen stores available in the lungs and have to rely solely on what is available in the bloodstream. When you get close to needing to gasp, you can extend your limits in two ways. The first is the same as with basic breathing, slowly letting out what is left of the air in your lungs. The second method will become critical later for controlling vasoconstric-tion. It consists of a rolling set of muscle contractions that you start at your feet and sequentially tighten until you reach up to your head.

The process is as follows: Relax your body and clench the muscles in your feet. Then clench your calves, then thighs. Work the contractions up your body until every part of you is tight from the bottom to the top. Clench your stomach, your chest, fingers, biceps, and jaw. Tighten the muscles behind your ears and imagine all of this pressure that you've built up going out the top of your head like you were rolling out pizza dough. Whenever I do this I end up making all sorts of grunting noises and squint my face into awkward contortions. It feels like I'm going to pop. But I never have. Once you finally have to breathe, take in a half lungful of air and hold it for about 10 to 15 seconds. This is the recovery breath, and it feels awesome. Now start over from the beginning. Since your lungs start near empty, it won't be possible to hold your breath as long as with the basic breathing technique. Aim to increase the amount that you hold your breath with each repetition. When I do it I start with a 1-minute hold, then 2 minutes, then 3. Even though everyone's physiology is different, Hof says that at 3 minutes you've cracked into your sympathetic nervous system.

That's the physical side of the exercise, but there's a mental routine that you can do alongside the breathing. One of the goals of this practice is to shut down the higher cognitive functions so that you can communicate directly with your lower brain functions. The brain burns far more energy than any other part of the body. In a typical day an adult male burns 2,200 calories, but the brain gobbles up 15 to 20 percent of that energy. To put this into context: If you charge up all 1,576 stairs from the ground floor to the observation deck of the Empire State Building, your muscles will burn only 54 calories, which is roughly the equivalent of a single Oreo cookie. (This is not to say that going up all those flights of stairs won't burn calories, but most of the additional energy actually stems from increased circulation and heart rate.) If you can turn down even a few of the brain's higher functions, you will see a vast improvement in your ability to hold your breath. And there is no better tool to dissect and modulate the inner workings of the mind than meditation.

People have been meditating for thousands of years. The very earliest written descriptions of meditation emerged in India almost 3,500 years ago. Oral traditions extend even further back. Yogis, monks, and spiritual seekers have developed hundreds, if not thousands, of meditative techniques aimed at probing the very innermost corners of the mind. The most basic, and, some might argue, most important, goal of meditation is simply to calm the mind and stem the ceaseless stream of thoughts that we have all day. It is possible to adapt just about any meditative practice that you might have with the Wim Hof Method, but the easiest ones tend to focus on visualizations.

VISUALIZATION MEDITATION

Even the blackest blackness behind your eyelids hides an array of colors and shapes. While there are centuries of explanations of what the shapes mean—or, indeed, if they mean anything at all—one thing we can be sure of is that when you close your eyes you still process information. This meditation aims to focus your thoughts so that your mind does not wander. Your mind will use less energy if it can focus on only one thing, and will allow you to go deeper into the physical experience of meditation.

Start by closing your eyes. Clear away any thoughts or mental clutter from your day. Don't muse over the bills you have to pay, the conversation at the office water cooler, or what a friend said the other day. Just sit in one place and try to be in the moment. Notice that you might be able to make out shapes in the blackness. Perhaps you see small flecks of reds, blues, and greens, or maybe there is a blotch of color floating out in space. If you don't see anything, don't worry about it, just look at the blackness itself. But, maybe it looks as if there are electric currents in the play of lights. Notice the outlines of the shapes and try to sense how deep the space in front of you really is. Is it a shallow blackness that stops at your eyelids, or is it a deep and spacious void? Simply observe the character of what you see. If you start to think about other things, let them go and gently guide your focus back to just looking at the shapes you see. Imagine that some of those shapes are actually light and try to feel that light entering your body. Let the light come in through your eyes or through your forehead and let it flow down your spine and out to your fingers and your toes. Keep focusing on the movement of the colors and start the power breathing.

Take 30 breaths while focusing on the light and as you begin to hold your breath. Notice how your body feels. Are you dizzy? Do your fingers or toes itch? Are you cold? Or hot? Feel your body and imagine that the light you can see stretches all the way to your toes and is actually air. Imagine that you're filling your body with light and with air as you hold your breath. Then, as holding back from breathing gets difficult, start to clench your muscles from your toes on upward and think that you are pushing air up to your head. Notice how it feels, and when you have to breathe, do it. Repeat this cycle of visualization until you can hold your breath for at least 3 minutes.

There are many explanations of what happens during meditation visualizations. Monks, yogis, and Chinese meditation practitioners say that the colors correspond to a system of spinning chakras (wheel-like torrents of energy) that make up a human's spiritual anatomy. Others argue that the colors indicate the metabolic functions of specific organs. Most scientists say there is no proof that the shapes a person sees are anything more than just the meaningless static of the brain. Whatever the truth of the matter, the visualizations are real experiences. When you give yourself a way to connect the visual information behind your eyes to the sensations in your extremities you have unlocked a tool for taking control of the process of the immune system itself. The combination of breathing exercises, muscle flexing, and visualization are the basic core exercises of Hof's training.

But meditation and breathing practice alone only comprise half the equation. While these activities can prime the nervous system for ceding some control to the conscious mind, in order to make real progress you need to use the environment to trigger an autonomic response that you usually have no ability to access on your own.

GETTING COLD

The first time that you try to stand in the snow it's not only going to feel cold, it's going to hurt. Even though it won't hurt any more than, say, a 5-mile run, or hitting the gym to lift weights if you're severely out of shape, the sense of fear that comes from the cold reaches down into some dark primordial core inside every one of us. The mere thought of it makes most people cringe. Your brain is probably screaming, "Hell no!" If you urge your friends to try it, their responses will resemble the various stages of grief: denial that it's a good idea, anger that you even suggested it, bargaining for some other exercise option, depression (okay, maybe not this one), and, finally, acceptance that it probably won't kill them. Almost everyone claims that while some people are able to stand in the snow, they themselves are especially vulnerable to the cold. The fact that this is a near-universal response might persuade them, but, perhaps, if you persevere, they might at least try it.

There is nothing special to the technique. You can fill a bucket or bathtub with ice and jump in. You can turn the shower on as cold as it gets and soak for a minute. Or, perhaps the closest to what our ancestors faced, wait for it to snow and then walk outside in nothing but shorts. Whatever the cold source, the goal is to give your system a little shock. Don't ease yourself into the frigid waters and wait for your body to acclimatize. Jump right in and see how your body responds, like you are doing a polar bear plunge.

Start with taking 30-second cold showers and build up from there. The immediate sensations are rarely pleasant. You'll start breathing fast, your pupils will dilate, and you'll want to start moving to keep yourself warm. Feet in the snow will often turn red as blood courses through the arteries before they turn lighter in color, once vasoconstriction sets in. While you're in this moment of shock and pain, you have two basic goals. First, you need to control your breathing and keep calm. The sensations that most closely match the feeling of burning will dissipate if you focus your mind on the pain. Try to relax instead of tightly clenching up every muscle. Enduring the snow is the same exact mental trick as trying to delay a sneeze or resist being ticklish. Once you feel reasonably calm, the second thing you need to master is suppressing your impulse to shiver. After the initial shock of the cold recedes, your body will soon click on an instinctual program to warm itself through muscle movement. It's up to you to keep that at bay. Shivering will certainly use up a lot of calories, and for this reason it can be useful if you want to lose weight. But your goal here is to command your nervous system to submit to your will. If your body can't shiver itself warm again or rely on the insulating properties of your white fat, its only remaining option is to start ramping up your metabolism. This is to say, you will begin passively generating brown fat and building up your stores of mitochondria if you can suppress your shiver response.

The first time you experience the cold will be the worst. Every nerve will fire like it is raw and has never been used before. Your heart will race and you'll want to do anything you can to retreat into warmth. Unless you are in weather where frostbite is likely, you're not in any serious danger, even though your mind will tell you that you are. The arteries in your hands and feet will clench with the force that it reserves

		Temperature (°F)																
Calm	40	35	30	25	20	15	10	5	0	-5	-10	-15	-20	-25	-30	-35	-40	-45
5	36	31	25	19	13	7	1	-5	-11	-16	-22	-28	-34	-40	-46	-52	-57	-63
10	34	27	21	15	9	3	-4	-10	-16	-22	-28	-35	-41	47	-53	-59	-66	-72
15	32	25	19	13	6	0	-7	-13	-19	-26	-32	-39	-45	-51	-58	-64	-71	-77
20	30	24	17	11	4	-2	-9	-15	-22	-29	-35	-42	-48	-55	-61	-68	-74	-81
25	29	23	16	9	3	-4	-11	-17	-24	-31	-37	-44	-51	-58	-64	-71	-78	-84
30	28	22	15	8	1	-5	-12	-19	-26	-33	-39	-46	-53	-60	-67	-73	-80	-87
35	28	21	14	7	0	-7	-14	-21	-27	-34	-41	-48	-55	-62	-69	-76	-82	-89
40	27	20	13	6	-1	-8	-15	-22	-29	-36	-43	-50	-57	-64	-71	-78	-84	-91
45	26	19	12	5	-2	-9	-16	-23	-30	-37	-44	-51	-58	-65	-72	-79	-86	-93
50	26	19	12	4	-3	-10	-17	-24	-31	-38	-45	-52	-60	-67	-74	-81	-88	-95
55	25	18	11	4	-3	-11	-18	-25	-32	-39	-46	-54	-61	-68	-75	-82	-89	-97
60	25	17	10	3	-4	-11	-19	-26	-33	-40	-48	-55	-62	-69	-76	-84	-91	-98

Wind (mph)

Frostbite Times: 30 minutes | 10 minutes | 5 minutes

$$\text{Wind Chill (°F)} = 35.74 + 0.6215T - 35.75(V^{0.16}) + 0.4275T(V^{0.16})$$

Where, T = Air Temperature (°F) V = Wind Speed (mph)   Effective 11/01/01

This chart will give you some established guidelines for safe cold exposure. Avoid frostbite, but don't worry too much about just feeling cold. That's mostly in your mind.

for severing a limb. If you are jumping into a cold body of water, go with a friend just in case one of you runs into trouble. Nonetheless, if you last just 5 minutes your entire body will probably be screaming for relief. Go inside and warm up. Most likely, the sensation of vasodilation, when the arteries loosen and let warm blood back inside, will hurt even more. For some, the process of rewarming is more painful than the initial exposure. Think of the pain as a sort of baptism. When you try it again the next day, it will be much more manageable.

Of course, not everyone has access to cold on demand. If you live in a temperate climate or are in the throes of summer, then you may have difficulty finding a reliable source of icy water. A lot of this is going to be up to your own ingenuity. Maybe you just take cold showers—turn the knob as low as it will go and let the cold water cascade over your head, back, chest, and legs. Or maybe you can spend an inordinate amount of time in the freezer section of your grocery store. Perhaps you fill up a bathtub with ice. One person I know who lives in the Caribbean devised a shower head out of a bucket of ice that slowly drips ice water onto his head. Another thing you can do is simply turn down the thermostat to below 66 degrees in the winter. That's the temperature that most people will feel the need to shiver if they're not moving around too much. By simply keeping the ambient air temperature that low (and not wearing lots of layers to insulate yourself from it), you will force your body to adopt a different strategy to stay warm. Whatever you end up doing, you need to get to a place where your ordinary response would be to start shivering, then do everything you can to resist the temptation.

ACTIVE CONDITIONING

Another very simple technique is what I call active conditioning. This is where you start whatever workout you normally do in one environmental extreme or another. I often will go running in just shorts and sneakers during the middle of winter, when snow has blanketed the Colorado landscape. I do a 3-mile run around a lake near my house without a shirt on while other brave souls are clad head to toe in fleece. Heads turn and most people will think I'm crazy, but anyone who has even a modest running habit knows that once they get moving their core temperature spikes upward. Those fleece-clad joggers are probably boiling up. However, since the nerves in your skin can still detect the weather, outdoor exercise in the cold while shirtless will send the appropriate signals to start generating brown fat. Shirtless runs end up being one of the most comfortable ways to practice the Wim Hof Method. If it's really cold, it's not cheating to wear a hat or gloves to ward off frostbite in the extremities.

The same thing goes for hot weather. When the mercury in the thermometer rockets upward, do a run in the middle of the noon-day heat. Don't push yourself so hard that you pass out from heatstroke, but get used to a slightly uncomfortable temperature. You can do the same thing in any weather, be it rain or shine. Part of the reason humans have gotten weaker as we've insulated ourselves with technology is because we don't experience variety in our daily lives. Make yourself a little uncomfortable, and all of a sudden you might discover that it isn't nearly as bad as you imagined it might be. One more thing: Remember to wear sunscreen.

BROWN FAT ACTIVATION

It seems to take about a week of training before your body recruits enough brown fat (BAT) for it to have a meaningful impact on your metabolism. If you don't live in a cold climate, it is more than likely that you are going to start out with near-zero BAT. Over time, however, your stores will grow, and that brown fat will passively suck white fat from your bloodstream and burn it for body heat. However, there are moments when you want more than just a little heat over a long period of time. What if you want to get warm right now? After a few weeks of training you can make BAT activate on command. The process for doing it is a familiar one. Once you're already cold, clench the muscles in your feet and hands and then sequentially tighten every muscle above them. The method is essentially the same as the one you use during power breathing except that now, rather than simply pushing all the blood up to your head in general, you instead focus on a specific point on a small ridge on the backside of your skull where the temporal bone and occipital bone meet just behind your ears. Once you've pushed all of your muscles up to that point, you then squinch any muscles you have in your scalp and head and pressure downward to the same spot. Essentially you're trying to pinch this area with muscle contractions. This is a somewhat advanced technique and the pressure could make a person pass out. So don't force it. Remember what happened to Wim in the sauna in Poland: Definitely don't try this when you're in a hot place. While it doesn't seem to work for everyone, many people will feel an intense burst of heat. For me, it sometimes arcs down my neck and into an area between my shoulder blades. Other people have described it as a fuse burning through their body. Either way, get ready for a lot of heat.

A 15-MINUTE DAILY WORKOUT

My daily routine mixes elements of all these practices, plus a few more I borrowed from a local yoga class. I always start before breakfast, when my circulatory system is the most active and hasn't yet been weighed down by the process of digestion. I begin with three rounds of power breathing followed by a breath-hold with no air in my lungs. I time myself and try to add a minute to each breath-hold until I hit 3 minutes. Then I do another round of breathing followed by 50 breath-out pushups. I follow this with a headstand for 30 seconds to let blood move to my head. I usually start my shower warm, but finish with at least a minute of cold water. I get out feeling refreshed and pumped full of endorphins. On top of that I do some kind of cardio exercise outside about three times a week. All together it really isn't very time consuming. It's also pretty much all I will do to prepare myself for the coming challenges.

5

BIB 2182

SOMETIME IN THE middle of July 2015 I take a drive into the Rocky Mountains to a ski resort that has a strategy to remain relevant during its offseason. Outside the base station there are a dozen merchants hawking athletic apparel, heart-rate monitoring systems, sneakers, and T-shirts. A few thousand people mill around the lodge sipping beer, chewing on cardboard-flavored energy bars, and waiting for their designated starting heat in an athletic event that is both totally mainstream and fairly new to American sports. Just above the parking lot where I am standing a series of rope ladders, mud crawls, water barriers, and god-knows-what-else are waiting for the masses to defeat them. The obstacle course, this one called a Spartan Race, is supposed to push me to the very limits of my mental and physical endurance. The race coordinators promise that I will suffer over the course of a few hours and become a better man for it. For some reason this sounds like fun.

So when I report to the registration tent with my wife (who enthusiastically said yes to the idea of participating), a 20-something woman gives us two small packets of vital racing paraphernalia and crosses off our names from a printed list. In my envelope is a radio frequency identification (or RFID) chip that I will need to wear in order to track my time over the course, a headband with my bib number, and a voucher for a free beer when it is all over. There is also a waiver to sign so that we can get past the gatekeeper. The most important part of the document stands out in bold red type so that those entering into the agreement will have no legitimate excuse to argue that they didn't see it just in case they decide to lawyer-up some time down the road. It reads: **"THERE IS A REAL POSSIBILITY THAT <u>YOU MAY DIE</u> OR BE CATASTROPHI- CALLY INJURED.** Each participant voluntarily and knowingly accepts and assumes all risks . . ."** The mandatory waiver gets more precise in the fine print, enumerating 14 different ways a person could end up in the hospital as a direct result of their participation in the event. The list includes drowning, near-drowning (which apparently needs to be specified), sprains, strains, fractures, burns, hypothermia, animal bites, poisoning, heart attacks, contact with fecal-contaminated water, permanent paralysis, and, yes, that run-of-the-mill catch-all: death.

The obstacle course race (or OCR for the people in the know) industry started as a single, intense mud run in England in 1987. It went mostly unnoticed in the world until 2009, when the idea of putting rope nets, barrier walls, ice patches, and electrified wires on a race course crossed the Atlantic and set up shop in North America. Today, OCRs reap a reliable $326 million every year from entrance fees and sponsor deals. The Spartan Race features a partnership with Reebok and a production deal with a reality TV show, but it's hardly the only game in town. Other races include Tough Mudder, Warrior Dash, BattleFrog, Strong Viking, Civilian Military Combine, and a seemingly endless list of spinoffs, competitors, and pretenders to the OCR throne. All of them share one thing in common, though: Together they've managed to persuade tens of millions of people that suffering across potentially life-threatening obstacles is a great way to spend a weekend. As the holder of bib number 2182 I'm hardly one to pass judgment.

The OCR industry is the fastest growing sport in the United States.

It has its roots in marathons, Ironman, and trail racing, but the bar to entry—or at least participating and earning a medal or T-shirt with the word "finisher" on it—is far lower. At least 54 million Americans call themselves "runners," which makes it the second most popular fitness activity next to walking. (That fact, pulled from a report by the Physical Activity Council, seems to suggest that the remaining 208 million Americans have truly sorry workout regimens, and probably helps explain why so many people in this country complain about their ever-expanding waistlines.)[1] In 2009, roughly 120,000 people lined up at 11 obstacle course races run by a company called Warrior Dash.

Tough Mudder started in Pennsylvania a year later. By 2014, 3.4 million people were willing to shell out entry fees between $100 and $200 to abuse their bodies on whatever challenges organizers threw at them. That the financial barrier to entry is so high probably also reflects the racial and sociological disparities in the country at large as well as in social media, through which most OCR participants reveal themselves to be young, white, and apparently possessed of some disposable income. Indeed, much of the trend has been fed by smart social media campaigns. Photographers posted at every major obstacle at the Spartan Race take tens of thousands of time-stamped pictures, often tagged to the same microchip that racers wear to keep track of their times. The pictures are free to use, which helps explain why just hours after a given race Facebook, Twitter, Instagram, and other channels explode with pictures of mud-covered friends vaulting over flames, enduring tear-gas clouds, or crawling on all fours through beds of snow. The appeal of temporary suffering attracts new racers and weekend warriors to the field of obstacle glory.

With the exception of elementary school romps over jungle gyms, I have no experience with this sort of thing. This is going to be my first

---

1   Quality studies of the general exercise habits of Americans are kind of hit-or-miss, and finding a convincing figure can be tricky. For instance, according to a study by Breakaway Research Group, 100 million Americans rode a bicycle at least once last year. Whether that makes them "bikers" or whether the "runners" identified in the other study cited here qualify for that designation if they went on one or more runs in the previous year is a judgment call you can make for yourself.

OCR, and thus it is my chance to discover what all the fuss is about. It seems to me that people who are up for running through mud must have a similar disregard for comfort as those who don't mind plunging into a tub full of ice. An obstacle course race also speaks to different mind-sets around fitness in general. Exercise doesn't have to mean plugging away on repetitions on scientifically calibrated weight machines, or running increasingly long distances on endless miles of roadways. OCRs favor sports dilettantes, for whom fitness means being skilled at all sorts of different athletic challenges as opposed to being a specialist in any one activity. Part of the allure also harks back to the time many of us spent playing as children, when climbing up a rope or crawling through a discarded tire was about as much fun as you could pack into a day. And, in a way, this brings me back to the waiver.

At first glance it's hard to tell how much the waiver is there for the entertainment of the participants—part of the bragging rights that anyone might claim after crawling under barbed wire—or whether it is actually here to disclose the very real possibility of death and dismemberment. For me, and probably most other people lined up in the morning heats, it's both. The threat is part of the allure. It is as if those of us who are lucky, skilled, or just tough enough to make it to the finish line do so knowing that someone else might not.

Every obstacle on every course breaks down into two basic challenges: the physical and the psychological. Carrying an 8-gallon bucket of dirt up a hill after a 5-mile run might leave your arms aching the next day, but there's no moment when you fear for your life.

Other challenges, however, stoke real terror. With races regularly employing tear gas, ice water, electricity, fire, and dizzying heights, the challenge of getting past them isn't usually physical, but is instead about breaking past mental barriers. The challenge is overcoming your natural instinct to curl up into a ball and wish for home. The popularity of these sorts of obstacles, in part, comes from their visual appeal, and they are especially striking in the photos on social networks after the race. It just makes you a little more badass at the end of the day if you can brag about what it feels like to breathe tear gas. In the era of an all-volunteer Army, obstacle course races are about as close to a military-style rite of passage as any of us are likely to experience. This is perhaps why so many OCRs borrow their names from military terminology and from warriors long past. Casualties aside, a lot of people want a chance to test their mettle in a way that could plausibly kill them.

There are at least a few studies that attempt to put numbers on the likelihood of the things mentioned in legalese actually happening. One, published in the journal *American College of Sports Medicine,* examined several Tough Mudder events in 2013 and found that 37 people out of every thousand ended up in the hospital on race day. There were injuries to the legs and joints, muscle tears, and broken bones. While minor injuries were almost routine on these courses, actual fatalities were still rather low. After consulting experts on the business, and a deft search of the internet, I am able to identify four deaths across the entire industry, which claimed to have 5.6 million participants in 2015. By contrast, skiing, which boasts 11 million downhillers, kills 40 Americans a year, on average. So yes, the races do indeed carry risks, but they're not extraordinary ones.

Injuries are a different matter. A case study in the *Annals of Emergency Medicine* surveyed one emergency room that was unexpectedly inundated by casualties. One young racer complained of chest pain after 13 consecutive electrical shocks, and a handful of contenders ended up in the emergency room incoherent from pain or exhaustion. One 31-year-old man may have suffered a seizure or stroke and was completely unable to move his right side after completing 20 out of 22 obstacles.

On a more intimate level, one that I have to admit most people probably aren't actively worrying over, there's something profound about facing down a situation that might trigger the fight-or-flight response. Be it a cloud of tear gas, a live wire, or the view from the top of some obscenely imposing jump into cold water, when the brain recognizes what it believes to be a legitimate threat, the sympathetic nervous system kicks into action. The brain's emergency signals will tell the adrenal gland to trigger a cascade of hormonal responses and release the neurotransmitters nor-epinephrine and epinephrine into the bloodstream. The heart rate goes up—often thumping audibly in your ears—digestion stops, the pupils dilate, and in extreme circumstances bladder control could be lost.

That all sounds bad, but the high you get from adrenaline is total. The world shifts sharply into focus, time seems to slow down, and your reactions quicken to a razor's edge. For the time that all that feel-good juice pumps through your bloodstream you are in such a heightened state of alertness that you feel superhuman. More often than not you stop thinking altogether, reducing the higher mental functions to the bare tools you need to survive in a given moment. Your body might be so focused on the task at hand that you don't even form coherent memories. But none of that matters; it's addictive, and awesome, and only happens when your body switches to survival mode.

Adrenaline doesn't provide an enduring high. Eventually the feeling of exultation descends into grogginess and fatigue. But the surge is so fundamentally mammalian that it is etched indelibly into our evolutionary past. From the perspective of someone who is trying to use the environment to trigger autonomic responses, OCRs represent a potentially potent tool that a cognizant athlete could use to train their fight-or-flight response and in so doing strengthen their sympathetic nervous system. The obstacles create a sort of safe stimulus to trigger some sort of neural response in the body. If you know what's coming, then a strong stimulus is also an opportunity to train yourself to resist and control potentially beneficial hormonal releases. In this sense, fear of a scary but nonlethal obstacle works the same way that the Wim Hof Method tricks the body by hyper-saturating the blood with oxygen (and blowing off CO_2) so that the mind loses track of its preset response for gasping. The sweet spot at an OCR would be finding a challenge where you are so overcome by an animal impulse for survival that you lose track of who you are in the moment and simply respond to the environment you're in. If you could condition that release, then it stands to reason that over time your adrenal response could be made to cave to your will.

Yet for the vast majority of people on the course today, there is a simpler reason to love obstacle racing. Most sporting events in America are competitions with clear winners and losers. One team overcomes another, or a super-athlete dominates a field. Today's OCR event is different. Outside of a few elite competitors, OCRs aren't about beating the guy next to you as much as they are about overcoming whatever the course throws at you. In other words, I didn't come to win. I, like millions of other contenders, came to finish.

A few minutes past noon I arrive at the starting line, where a man with a microphone and Army-camo shorts urges would-be Spartans to crowd together in a bullpen in front of an inflatable race gate. To get there a hundred of us have to surmount our first obstacle: a 4-foot wall made of plywood and 2 x 4s. It's just a taste of what is to come.

"When I ask who you are, you yell back, 'I am a Spartan!'" he calls into his microphone. It is the 50th heat of the day, but the emcee doesn't appear tired of the routine. His lips peel back into a menacing grin as he asks his question. And once we call back he starts chanting the Spartan war cry, "Aroo! Arooo! Arooooo!"—the Greek version of the Marine Corps "Ooh-rah." My wife rolls her eyes at the regalia and asks me to promise that I won't play along. Even though the obstacle course industry is reasonably gender balanced with four women entering for every six men, the overall feeling of most races brims over with machismo. I "Aroo" anyway. We are Spartans, after all.

Then there is a puff of smoke from a firework and the slow rumbling of the legion of would-be Spartans in Spandex running shorts, headbands, and nervous energy pushes forward up a green grass slope that serves as a ski hill in the winter months. My wife and I bring up the middle. For the first hundred yards or so the stampede moves like it is shot out of a cannon, but as the grade of the hill takes a steeper turn the crowd slows to a crawl. For a 6-mile race called a "sprint" it is hard for me to tell if this is just pacing in anticipation of some really grueling obstacles that the pack-leaders know about all too well, or if instead it is a sign that none of us in this heat are actually extreme athletes.

The course weaves in and out of an alpine landscape, and the moment the mountain levels off, the trail crashes into a shallow and slightly chilly pond. We dive in head-first, then surge forward out of the muck until we come to a giant pile of snow that has been transported here from higher up on the mountain. This part of the course makes us crawl beneath barbed wire, and we scrape our knees on jagged edges of packed ice. Blood drips down my legs. The sight of the superficial wound invigorates me for more. I get up and begin a dash downhill, where there are a series of variously tall walls to leap or scramble over. By mile 5 I am not nearly as spent as I'd hoped to be. I'm beginning to wonder if I should have signed up for the alternate, a 16-mile course called "the Beast" that adds more obstacles and another few thousand feet of elevation gain.

But the race isn't over. We jog forward with a thousand other people until the course intersects with an area where very tired-looking contenders are on the final leg of the Beast. They've done an additional 10 miles more than what we have and look the part. Their faces are drawn and white with exhaustion and their paces are measurably slower than us sprinters. I'm a little jealous. The courses join in front of an 8-foot boundary wall that is now a choke point for about 50 participants waiting their turn for a chance to hop over. Directly in front of me is a platoon of off-duty soldiers who have come up from a base near Denver. I watch as one shaven-headed lad with acne, no shirt, and blood streaming down his shins from an earlier fall runs full bore at the wall and jumps. His hands grasp the black-painted 2 x 4 on the top as his near-naked body slams against the barrier. He manages to half pull himself up before he runs out of juice and sinks to the ground. He looks down at his feet and runs back to the line, all the while apologizing to his company. "Sarge will never forgive me if I don't get over this," he says, determined to keep at it or die trying. So I wait and watch as he charges and face plants into the plywood a second time. On his third attempt his buddy gives him a helping push over. I hear him crash down the other side, no doubt happy to be done with this part.

Unlike traditional races, obstacle courses don't favor specialists. The variety of challenges play to different strengths, so very few participants excel at everything. The lean build of an ultra-runner may help with the steep hill, but doesn't play favorably with a 20-foot rope climb that requires upper-body strength. Fear of heights could stymie someone at the top of a wall, while a great jumper might have too much muscle weight to really compete with someone who trains for marathons. In this sense, obstacle courses bring out the best and the worst in an athlete's training. The people who make it into the elite tier are well-rounded, but they are not likely to excel in a traditional pro sport. No doubt the soldier who barely made it over the wall will crush the next challenge just around the corner.

After my wife disappears on the other side of the bulwark, I make a short sprint and, despite my pendulous calf muscles and a 6-foot-2 frame, easily grab the top of the barrier that had stymied the soldier. Still flush with momentum, I sling my foot sideways to latch on to the top of the wall and begin to haul the rest of my body over. Then, somewhere in the middle of that movement, I feel a barely audible pop in the vicinity of my pelvis. When my feet make contact with the soft ground on the other side, spears of shooting pain course directly into my brain. I walk a few steps and stop in my tracks. My wife runs in place next to me and gives me a worried look. I press my hand into the part of my groin where the ball head of my femur attaches to the hip socket, and I am pretty sure that something isn't quite right. But I am not ready to hang up my racing laurels, either. In the obstacle course world, a DNF—for "did not finish"—is a mark of shame. So, rather than submit to the pain, I limp forward a few more paces and grit my teeth. "I can do this," I tell myself, figuring that if I just keep moving that the joint won't have an immediate excuse to stiffen.

My wife and I inch gingerly up a hill at a slow jog until we pass a woman in pigtails, a neon green headband, yoga pants, and a compression bandage wrapped tightly around her ankle. She is moving at a crawl, but she is still moving. I ask her what happened. "I sprained it back there," she replies through a grimace, adding, "but I'm going to finish this. There's no way I'm going to be carried off the field." I offer to help, but she wants none of it.

The course opens up again a few dozen feet later and now I can see hundreds of people on a long straight stretch. At least a quarter of them jerk and scowl as they favor one limb or another. Pain is everywhere. And in a way, seeing all that suffering is part of the plan in the first place. No one starts out wanting to get hurt, but even though we don't like it, we all share the experience together.

A handful of obstacles later I find myself near the end of the race, climbing a 20-foot rope hand-over-hand above a muddy pool of ice water. The goal is to ring a cowbell at the top, but three-quarters of the way up I look down. It is a mistake. The world seems to go off-kilter and I imagine what would happen if I let go and fell backward, maybe smacking my head and sinking below the opaque surface. Though I still have strength left, the vision of falling makes me pause. I slink down the rope, defeated not by my muscle, but by my mind. The punishment for not completing an obstacle in the Spartan race is 30 burpees, so I get down in the mud nearby and do my penance. Put off by the failure I slog on toward the last obstacle: a wall of fire. There is no challenge to speak of, just a small hop through a flame. It is over in a flash. I end the race triumphant.

Across the finish are more canned photos, a medal that certifies me as a bona fide obstacle course race finisher, and an invitation to visit a beer hall with anyone else who has just gotten off the course. Muddied and bruised, my wife and I find an encampment of hoses and we spray each other down with ice cold water, a final shock to the nervous system. I am tired, but not exhausted; amped up, but not exhilarated. And at least part of me still wants something more.

In appealing to the masses, the race promises anyone a chance to meet their limits—which maybe I did at the top of a rope with an un-rung cowbell—but no challenge pushed me to a razor's edge, where my mind shut off and the reflexes of my lizard brain took over. I'd hoped to corner my inner animal by finding the very limit of my endurance. I will carry the pelvic injury with me for a matter of days, but experiencing an injury was never the point. I wanted to be scared and exhilarated all at once, but maybe I'd just chosen the wrong event. It wasn't necessarily the Spartan Race's fault that I hadn't opted for the longer course. There are hundreds of obstacle course races every year that might yet leave me even more burned, battered, and bruised in the end. I've already received an invitation from a friend who is making a film about obstacle course races to visit a course in England, sometime in the depths of the next winter, that is reputed to be the most brutal of them all. It's an appealing opportunity, and one that I plan to take him up on. However, I suspect that before I get there I need to find another way to capture that moment of primal awareness that I'm looking for. I need a different way to break into the deepest recesses of my brain. And so I decide that I'll need to find someone whose pursuit of adrenal bliss puts him in constant contact with his own mortality. A man who has no liability waiver to serve up.

6

ART OF THE CRASH

ON AUGUST 7, 2000, on a shallow reef off the coast of Tahiti named Teahupo'o, or "place of skulls" in the local tongue, surfing legend Laird Hamilton clung to a towline behind a Jet Ski piloted by his friend Darrick Doerner. The swells that day were slow, lumbering giants that seemed to drag the full weight of the South Pacific Ocean with them. The waves at Teahupo'o reef form deep in an endless abyss inside an oceanic trench before rocketing upward to a point as shallow as 20 inches below the surface. The sudden change in depth funnels the water in such a way that not only results in truly monstrous waves but a regular, semicircular presentation with an almost perfect hollow break. It's the sort of ideal spot that surfers have always dreamed of exploring, but there was no practical way to paddle over the monsters without risking life and limb. Hamilton's idea was to circumvent the most obvious obstacle by strapping his feet to a surfboard and asking his friend to tow him into the wave behind a Jet Ski. Until this point, no one had ever attempted anything like it. Nervous, Doerner and Hamilton watched the ocean focus its power on one particularly imposing swell. Doerner gunned the engine and brought Hamilton screaming to the precipice of the rising water. As it began to curl, Doerner had second thoughts. He wanted to turn back to tell Hamilton to abort. The water was just too big. The swell they were cresting wasn't just any wave, it was more like a tsunami. But when Doerner looked for his partner, Hamilton was gone.

After letting go of the towline, Hamilton felt the immense power of the water below him, and his board sucked backward in a vortex of fluid that seemed to twist the very logic of gravity. The upward force of the water was so strong that he had to steady himself by dipping his hand into the water on the front side of the board—a movement that most surfers who fear falling over the leading edge of their boards find antithetical to balance. Yet to keep upright Hamilton had to account for the unexpected current. As the wave crashed around him a videographer in the water captured his image as he disappeared into the collapsing tube. Several agonizing seconds passed. Doerner's heart sank as he imagined Hamilton's battered body torn apart by the crushing water. He was sure this blur of white water and deadly sea foam had just ended his friend's life. Then, inexplicably, Hamilton shot out of the pipe intact, jubilant, and humbled.

With a face at least 40 feet high and more than 100 meters long, it was the heaviest wave that had ever been ridden, estimated at about 30,000 tons. Papers dubbed it "the millennium wave." *Surfer* magazine ran the photograph of Hamilton crouched low on his board with a headline so flabbergasted that the editor couldn't even muster capital letters, just the words "oh my god . . ."

The video of the ride ushered in a new era of tow surfing. The YouTube video has racked up 1.1 million views, while a more recent video of another surfer on the same wave drew 26 million sets of eyes. It's the sort of feat that only the very best watermen can even attempt. One where only the lucky survive. Since 2000, at least five surfers have died trying to replicate Hamilton's effort.

Despite all of this, I've never heard of Hamilton until I begin looking deeper into practical applications for the Wim Hof Method. After the Spartan Race I want to meet a person whose connection to the natural world allows him to accomplish feats that are on such a grand scale that they are superhuman.

I call Scott Keneally, a journalist and filmmaker outside of San Francisco who has covered obstacle course racing and extreme sports for years. I tell him about my predicament and he can't help from laughing at me over the phone.

"You mean you don't know Laird Hamilton? Literally one of the best surfers of all time, does Wim Hof's breathing every single day?" he says, chiding me for my ignorance. "Hang on, I'll put you in touch. He doesn't use e-mail, so you're gonna have to call."

Over the next few days I read up on Hamilton's career. I call buddies who spend every opportunity on California's endless beaches. They can't help but relay their own chance encounters with him. One of them, a creative executive in an ad agency in LA, remembers sitting on the sand while eyeballing Hamilton as he "shot the pier"—riding a wave through the narrow confines of barnacle-encrusted support pillars where one inch of miscalculation would mean being caught between 50 tons of water and an implacable piling. Another friend remembers seeing someone on a stand-up paddleboard who was just a speck on the horizon but moving at an incomprehensibly fast pace. "It could only have been Laird," he tells me. The stories of Hamilton only grow with each telling. It is as if Hamilton isn't a person, but a folk hero on the order of Paul Bunyan. He is an adventurer who matches the likes of the great polar explorer Ernest Shackleton and George Mallory, the first man to attempt to summit Everest.

At the same time, Hamilton is remarkably accessible. Once I manage to get him on the phone I merely mention my interest in Wim Hof and any airs that might have been stewing vanish. Hamilton splits his time between homes in Hawaii and on the Southern California coast in Malibu. It so happens that he's developed a training program based in part on Wim's breathing techniques. He suggests that I book the next ticket I can find and come out to see him.

So a few weeks later I hop a plane to Los Angeles during one of the most severe droughts to ever blanket the region. The heat wave comes on the back of hot Santa Ana winds that blow in from Mexico. Just 10 minutes after arriving, the sweltering heat is enough to leave me drenched in sweat. I stay with a friend of mine in Long Beach, which is on the far southern rim of Los Angeles County. The drive to the most northern reaches of Malibu will take me clear up the coast, across the bustling center of the metropolis, and into the rustic beaches of the north. I am supposed to show up for the beginning of the workout at around 8 o'clock in the morning. But even though I make an early start, getting behind the wheel of a sub-compact rented Nissan before 6 a.m., I still find myself snarled in the city's impenetrable gridlock. As I inch up the 405's expanse of asphalt, I consult my phone for any hope of a shortcut. It suggests a 20-minute detour onto equally impregnable city streets before dumping me back out on the highway just a few blocks ahead of where I'd just been. If I'd consulted a map I would have known it was a hopeless effort to begin with. Unlike the ancestral wave pilots of the South Pacific, or even myself before the year 2000, I still have no innate sense of direction. It is well after nine when I show up in Malibu.

From the street, Hamilton's place is disguised by a concrete wall and a dull, gray iron gate. But the curb overflows with luxury automobiles, a dead giveaway that behind this barrier Hollywood glitterati meet for secret training sessions. I squeeze my rental between a Porsche and a Bentley. I approach on foot, and as I get closer I can make out sounds of splashing water and tribal music. I peek through a section of iron mesh and wave at a man in a swim cap for the gate code. He yells back four digits and I let myself in. It is only when I get a little closer that I realized that the square-jawed gentleman is the actor John C. McGinley—a regular of the medical sitcom *Scrubs,* a veteran of the Oliver Stone film *Platoon,* and a hundred other movies and TV shows. Somewhere in the pool Orlando Bloom is swimming underwater laps while carrying two 35-pound weights. It is hard to make sense of exactly what this all means.

As it turns out, Hamilton owns one of the largest private outdoor pools in Southern California—which says quite a bit considering how access to water in Los Angeles is as much a status symbol as it is an occasional necessity to seeking relief from the summer heat. It's 12 feet deep in order to facilitate the underwater exercise routines he's been experimenting with. The boot camp that Hamilton runs, what he calls XPT, which stands for Extreme Pool Training or, depending who you ask, also sometimes Exploration Performance Training, is the most exclusive workout in the city. Here is where the music producer Rick Rubin—the cofounder of Def Jam Records and the producer behind hit albums by the Red Hot Chili Peppers, Beastie Boys, and Lady Gaga—lost 130 pounds with a combination of ice baths and outdoor recreation. It's where some founders of CrossFit explore new routines, and, most important, it's where Hamilton spends his off-season before chasing big waves throughout the winter.

At 6 foot 2 inches and with a sandy shock of blond hair that sweeps across his forehead, Laird Hamilton has the rugged good looks that betray how easy it was for him to make a living as a model in his teen years. Now in his early fifties, his body remains a wall of solid muscle.

"It's important to have an off-season," Hamilton tells me once I make it down to the pool deck and notice the conspicuous lack of surfing gear lying around. "It lets me anticipate what's to come."

Rather than spend a lot of time asking him questions, Hamilton tells me to just get in the water and get a feel for it myself.[1] He wants to put me through a couple of his introductory routines, so I dive in with a mask on and start treading water. Hamilton grabs two 25-pound weights, places them on the rim of the pool, and then jumps in next to me. Using only hand signals he directs me to watch his movements, and without more explanation he grabs both weights off the deck and sinks quickly to the bottom. Landing in an underwater squat he raises the two dumbbells over his head and then uses his coiled legs to blast off the

---

1   The underwater exercises described here are inherently dangerous and should never be undertaken without appropriate experience, training, fitness level, doctor approval, and supervision.

ground, sending his body rocketing toward the surface in a reverse dive. To add power to the movement he strokes downward with the 50 added pounds of lead, using them like makeshift fins. His head breaks out of the surface just long enough to take a giant breath of air before sinking again to the bottom. This is one rep of potentially dozens. The underwater power jumps are supposed to roll out one after another just like the pushups after Hof's breathing exercise. Once he's muscled through 10 repetitions he gives me a smile and places the metal weights on the pool deck. "The trick is in the jump," he says. "You won't get very much power in the stroke when you're holding the weights." And with that, I grab the same set of dumbbells and let them carry me into the depths.

At first the movement is not completely intuitive. It's strange to move underwater and not be swimming, but it's also impossible to float while carrying so many extra pounds. I look up though the blue waters and see Orlando Bloom swim a lap while carrying what looks like a 40-pound weight close to his chest. He is skimming along the surface using only his legs with the weight held firmly in both hands. It's a move they call an Ammo Box, and I am doing my best not to be starstruck while weighted down on the pool floor. So I focus on imitating Hamilton's form and plant my feet firmly on the ground, crouch into a squat, and launch myself upward while letting my arms fall to my side in a rough imitation of what I'd just seen Hamilton do. I have enough force to clear the surface and just enough time to gasp for air before the weights drag me back down.

The first few reps aren't too bad. I find a rhythm when I bob to the surface, take a gulp of air, and sink back down. But every time I hit the apogee of the movement I suck in just slightly less air than I've used. I'm running on a deficit. After five reps I begin to struggle. I am not sure I have the power to reach the surface with the lead in my hands, so I gather myself by walking along the floor of the pool to its concrete edge. I look up through the mask and figure I can make one last jump. I jump, push the weights above my head, and have just enough power to heave them up and set them on the deck. I catch my breath while holding on to the edge and look down into the depths at a hubbub of activity. John McGinley is walking back and forth underwater with a single 50-pound dumbbell clutched in both arms, only coming up for air after he is done walking two lateral laps.

Over the next couple of hours Hamilton runs me through a half-dozen other exercises. There is one called the Seahorse, in which I cinch 30 pounds of iron between my legs and try to remain in a seated position while also awkwardly attempting to perform an underwater lap with a sort of winglike breaststroke. There is an exercise sort of like the Ammo Box that Bloom had been doing, but with a weight so heavy that there is no chance of staying on the surface. Instead, the lap becomes a sort of descending line to 12 feet down to the other side of the pool. The exercise ends with a power jump back to the surface and a return lap with the same diagonal descent.

There are also just simple underwater laps in which I try to do as many lengths as possible on a single breath, or, occasionally, just one gulp of air on either side. Every exercise starts off reasonably easy but gets progressively harder as oxygen becomes a scarcer resource. It's the sort of training that works best in the company of world-class swimmers who are paying attention to every awkward stroke. The dangers, after all, are real. Getting someone out of the pool quickly if they "go floppy," as Hamilton calls it, is something they always have to be ready for.

Of the people at the pool that day, though, only one person has a reputation for consistently pushing things too far. After I swim until my muscles scream for a rest, I sit on the side to recover. Hamilton points down to Orlando Bloom, who is still swimming along the bottom of the pool in goggles and a blue latex cap. There is some obscenely large piece of metal in his hand. "We always have to keep a special eye on Bloom. Just about every time he's up here he goes floppy at some point in the routine," he says, adding that the episodes are so frequent that they have their own nickname: "We call them Bloomouts."

It's sort of funny, but there's a real potential for danger. If someone is trying these particular exercises without supervision, it would be pretty easy to reach that limit of human endurance and simply pass out in the water. While that might be okay with landlubber's breathwork, the consequences are more severe underwater where a gasp means drowning. But if I can get beyond that—and under Hamilton's gaze, I am hopeful that I can—then perhaps I'll be able to force the wedge a little deeper between my mind and body.

In any event, the weight exercises are surprisingly low impact. Most people leave Hamilton's workouts refreshed rather than exhausted. Part of the reason could be because he demands that everyone who attends also conditions their body with alternating ice baths and sauna sessions.

He shows me a hidden alcove between the pool and his mansion where he keeps an outdoor ice bath and his ice machine. "You have to stay in for at least three minutes," he says, gesturing to the aluminum contraption that looks like a cross between a cattle trough and an industrial milk tub. So I get in and feel the frigid water close around me. As I cool in the ice, I eye some sort of churning device at the end of the tub. When it's on, it keeps the water in constant motion. The moving water prevents body heat from forming a small, insulating layer of warmish water when a person remains still. It would be much colder if he were to turn it on. "I had the illusionist David Blaine out here a few months ago," Hamilton says. "He said that he could last fifteen minutes in the ice with no problem, but I turned on the circulator and he had to jump out in two minutes." I lay back and close my eyes. The ice feels good. I stay put for 5 minutes, and when I turn it on, the circulator changes the calculus of ice bathing. It takes more focus to stay warm.

Hamilton heard about the Wim Hof Method through a 10-week online course on Hof's Innerfire website. For Hamilton, using the cold makes sense. He grew up swimming in the tropical waters in Oahu, hypothermia was always a present danger. Even warm Pacific waters slowly leach away heat from submerged swimmers, and for people waiting around on a surfboard for just the right set without a wetsuit, it's easy for them to end up shivering uncontrollably. Hypothermia is far from unknown to warm water surfers. Yet Hamilton can spend half a day in the water and says controlling his temperature is half the reason he has any success riding big waves.

"One thing about waves and the cold is that there's no discrimination, no mercy or grudges. There are no opinions or intent when they take their toll on a person. They just do what they do. They will never surprise you, and their baselines never really change. They're forces of nature," he says. The forces he's contending with aren't prone to the whims of human irregularity, nor is there a persona that guides how they react. Nature doesn't care about how you feel about it. When a person's temperature drops to the point where their brain becomes cloudy and decision-making sputters, the ocean will not relent. Yet it has no malice when it takes a life. Hof's training represents a way to counter that indifference, and with it Hamilton can stay warmer in the water.

As a constant tweaker of workout routines, Hamilton has also found a few ways to eke out just a little more performance from the Wim Hof Method. While ice and sauna immersion between underwater routines improve the circulatory muscles in his vascular system, Hamilton has modified the breathing method to help him excel at endurance events on land. The way he explains it, every workout has an ultimate limit where the physical load of an exercise increases the heart rate, and accordingly respiration goes up to compensate until the body simply runs out of oxygen and the movement stops. This basic concept of biomechanics dictates that as the body moves, a person needs more oxygen to fuel the motion.

Exhaustion is just the inevitable result of what happens when breathing hits its maximum rate and the body can't get enough energy into the system. This is what physical trainers call VO_2 max. When you hit this moment of peak performance it's simply not possible for you to push any further. We've all been at this point some time or another. Your face is flushed red and your chest is heaving, but try as you might you know that in a second or two you're going to crash. Peak performance at VO_2 max is partly based on genetics, as everyone has a slightly different ability to metabolize oxygen. Yet Hamilton assures me that exhaustion like this isn't only a mechanical problem, it's also a conceptual one. "If you have to stop to catch your breath, then you probably started the race all wrong in the first place," he tells me. "Instead, if you lead with a breathing pattern like you are already at your maximum load, then you won't have to work so hard later."

He tells me to think of it this way: The body has an internal program that automatically evaluates and responds to the conditions it is working in, but it doesn't have any way to know on its own what sort of oxygen load it is going to need in the future. It responds to the workload of the moment. When a sudden spike in heavy exertion happens, the body has to play catch-up. When you don't plan for breathing in a workout, it's easy to end up with an oxygen deficit.

The solution is to start breathing hard long before you actually feel the need to. The most obvious application for this technique is with running. A typical adult takes about 15 breaths every minute when they are at rest, but will start breathing 40 to 50 breaths per minute during periods of intensive exercise until they hit their VO_2 max. An untrained runner will breathe faster as they reach their limit, but they typically draw shallow breaths that don't give their lungs a chance to maximize oxygen absorption. By anticipating what the body is going to do, Hamilton says anyone can tweak their physiology for better performance. To demonstrate, Hamilton shows me the breathing pattern he uses to lead into his workouts.

His lips bulge stiffly outward and his eyes narrow to slits as he sucks air in coarsely through his nose. Waiting just long enough for his lungs to fully draft, he exhales with a growl reminiscent of Maori warriors as they perform a haka. Taking a breath every second makes him sound like he is under heavy labor. His face flushes with exertion but his eyes have the serene gaze of a mind focusing only on the action of its lungs. After 50 breaths he inhales deeply and is ready to run. The process scrubs carbon dioxide from his lungs, allowing his body to run more efficiently for a short sprint in the same way the Hof breathing method allows someone to do more pushups than they expect, or, if he were to keep it up, throughout the duration of a longer exertion.

Hamilton begins every morning with 20 minutes of this sort of breathing. He says it helps him get ready for the day and clears his mind for the tasks that he has in front of him. Once he's in the pool—usually by 7:30—he does a quick round of breathing a few minutes before starting underwater laps.

Technically, hyperventilating and swimming underwater can pose a serious danger of blacking out and drowning, even for expert swimmers. Since the body is designed to sense CO_2 levels in the blood, not the presence of oxygen, a breath-hold diver who has expelled CO_2 out of his lungs has very little warning when his actual oxygen levels are seriously low. Freedivers generally stay away from the technique for their deep water plunges, and the U.S. *Navy Diving Manual* issues stern warnings to any military personnel who attempt it without proper safety protocol in place.

Which is why, when I join Hamilton in a series of breath-hold laps in his pool with three of his friends for the following morning session, I am wary of what might happen if I or someone else goes "floppy." To his credit, Hamilton recognizes that what we are doing today carries a risk of blacking out in the shallow end of his pool. So we are going to have to stay close together and be hyper-aware of how everyone else is doing in case someone needs to quickly exit the pool. Today's session is an experiment, and one that he is not sure will get incorporated into the canon of XPT.

The five of us face off on alternate sides of the pool's shallow end, with three on one side and two on the other. The plan is to do three rounds of breathing and breath-holds and then dive under the water with empty lungs to do as many laps as we can. "There's something different about swimming with no air in your lungs," says Hamilton. "You might not go as far, but you pass out in a different way. It's slower." Perhaps this is because the goal of swimming with empty lungs isn't about extending the length of time underwater as much as it is about expanding the wedge between your conscious mind and your body's actual limits. Since the body will process oxygen in the bloodstream and then expel the waste CO_2 back into the lungs, the body may have a chance to sense CO_2 build-up before actually blacking out. In Wim's terminology, breath-out exercises primarily work on the sympathetic nervous system—the fight-or-flight responses. Breath-in exercises first trigger parasympathetic responses and might make a person more inclined to relax before the sympathetic responses kick in.

We swim between each other like interlocking fingers and I easily reach the opposite side and flip around for a return lap. I remember a warning Hamilton gave me earlier to try not to sprint underwater but instead to take every stroke slowly and conserve my resources. I kick out to the alternate wall just a little slower than my first pool length. I can see that the others are headed for a third lap so I plunge ahead with them. Hamilton's blue bathing suit flickers past with fishlike efficiency. Three lengths down, I turn and head into the fourth lap with darkening senses. Four strokes later I can feel something shaking deep within me. I've been underwater for about a minute and I want to breathe, but instead I let out some of the CO_2 that has built up in my lungs. It is enough to enable me to finish the lap, but when I come to the surface the world feels a little darker than it had a minute or two earlier.

I grasp the edge of the pool and close my eyes. I take a single deep breath and then hold it. The shake that is welling up in me grows a little stronger and soon a shudder rocks through my core. Then, as I gaze out into the starry blackness behind my lids, I see a sunburst of reds and yellows. I can feel the pool tile beneath my forearms as my head bobs down. The fireworks hang in the black void for a moment and begin to transform into a shape that resembles a human face with two black slits for eyes and a halo of flame around them. The image stays for a second before I take another breath and open my eyes.

Darien Olien, a nutritionist who is a regular at Hamilton's pool, completes two more laps after I came up for air. He is breathing hard and has his arm around Hamilton's steely-jawed nephew—a professional stand-up paddleboarder from San Diego. They both hunch against the wall. All of us pushed ourselves to the edge and slipped into some sort of exercise-induced transcendence.

It's not uncommon to hallucinate during longer breath-holds. The deprivation of oxygen peels back layers of the brain, shutting them down until you reach a sort of primal center of human existence. The brain's electrical patterns fire in sporadic ways as a person begins to lose consciousness. The process is rooted in our basic biology, but for Hamilton the experience has deeper meaning than a simple physical reaction.

"Breathing lets you go deep," he says to me later, describing how his 20-minute morning routine of deep breathing and breath-holding has led him to places he never expected he would reach. "Sometimes I'll see a face looking back at me and I know that it's my own soul. At other times it feels like I'm floating above my body looking down at myself meditating." The face he sees most, he says, resembles the crude features of the comic book hero the Silver Surfer—something that is otherworldly and yet also human. I'm not entirely surprised that the deity he identifies with owns a surfboard.

It is open to debate whether what I've seen on the edge of unconsciousness is a connection to something that truly transcends the world we live in or is instead the machinations of my hypoxic mind attempting to create meaning out of meaningless physical phenomenon. Clinical literature often reports different types of hallucinations that can occur during meditation, and in most cases the experiences don't correlate with an underlying mental health condition. In India and China, yogis who use similar breathing techniques, called *pranayama* in Sanskrit, or *qi* breathing in Mandarin, frequently report experiencing visions. In fact, most Eastern traditions believe that breath control can engender profound spiritual insights. For Hamilton and, indeed, many people who delve deeply into their own physiology through meditation or yoga, there is an intrinsic connection between the body and the soul, and every such experience has the potential to open a window into something greater. Whatever it means, just like a dream the experience itself is real.

Also real, of course, is the possibility of actually passing out underwater. If I'd gone a little farther—another lateral lap of the pool—the vision could have been a precursor to my mind shutting down and fainting. In unconsciousness my autonomic nervous system would take over and my lungs would have filled with water. Shortly after the session was over Hamilton gets in touch with me to say that they aren't going to mix breathwork and waterwork anymore. While the underwater laps are interesting, the risks that come along with them just aren't worth it. He thinks it will be better to keep that part of the training on dry land.

Whether or not Hamilton needs to find out the ultimate truth of what happens to us after we die, I suspect that his faith in something greater than himself has helped him get through the potentially fatal situations he finds himself in during his adventures on the water. Peek at any of the dozens of videos online of a pea-sized Hamilton being towed onto the apex of a rogue wave and disappearing down its vertical walls, then try to visualize your own feet on his surfboard. The crushing energy of that water exposes everything in its path to the gargantuan forces of nature, in the face of which all of a human's dreams, goals, and ambitions are insignificant. The wave doesn't care if you live or die. Accepting the sheer mercilessness of nature is the key to riding a big wave in the first place.

A few hours after our pool routine, I sit with Hamilton outside his Malibu Hills home overlooking the ocean. He squints his eyes at a distant storm cloud as I ask him about how he faces the probability of death before dropping into a wave.

"You ride a wave, you don't conquer it," Hamilton says. "The trick to surviving is knowing when you're out-matched." In part, he credits his achievements as a surfer to what is for him a humble connection to nature and the realization that he is not just the rider of a wave, but indeed, that a surfer is *part* of the wave. "There's no beginning and no end of a wave," Hamilton says. "Instead there is a sort of timeless aspect to surfing where every wave you're ever on is part of the same eternal wave. Really good surfers have a connection to the ocean where the great waves simply come to them. Surfers who fight the ocean and try to defeat it never connect and aren't given good waves."

Even so, it might not be fair to say that the five people who have died on Teahupoo since Hamilton rode the millennium wave weren't connected with the ocean. Wipeouts and injuries are so common in his business that Hamilton stopped counting the stitches he's had sewn into his flesh after they surpassed a thousand. He's broken his ribs and his fingers, blown out his eardrums, and had the tip of an errant surfboard plunge through the flesh on his cheek. All this is to say that he's no stranger to the consequences of failure. The key to success for Hamilton is embracing the possibility of crashing. Or even death. Everything he tries starts out with a plan of what failure might look like. It's a recipe, he says, that applies to a lot more than just surfing. I ask him to outline what that means.

Before he drops into the wave, he envisions a successful ride. He thinks about how he will enter in at just the right angle and, no matter what the forces are against him, find a way through it. Once he has that plan in mind, he says, "I commit and believe that I will make it." That simple belief is often enough to see him through to the safety of the shore. But even if it doesn't, this attitude of success can pull him forward when things go wrong.

In the split second that he realizes that he's lost control, he thinks about what the crash might be like and then plans to minimize the damage. Whether this means rolling into a ball, diving off the board, or whatever else opportunity puts on his doorstep, he believes that there will always be a moment when he will have a semblance of control again. The trick is waiting for it to happen. The swirling currents and crushing forces are, after all, just forces. They ebb and flow and dispassionately dispense opportunities as they do obstacles. During a crash, the water rushes over his body and asserts its will. In the torrent of water his limbs aren't his own. He'll be smashed by water, rocks, the board he was carrying, and whatever else gets in his way. In these moments of utter powerlessness Hamilton submits and relaxes. In the eye of the maelstrom there's nothing to do but let nature run its course. It will either kill him, maim him, or let him go.

The Zen during the tumble is humbling and can tick by for breathless minutes. Hamilton's lungs might burn and his body might be brutalized by the swirl around him. "But at some point," he says, "if you make it far enough, the grip will loosen. You have to keep your mind still while you are in the midst of it all, knowing that your chance will come." And when it does let up, Hamilton taps into all of the stored up energy in his body to make his move. That's when he lets the fight-or-flight response take over. He uncoils like a jack-in-the-box and sprints with every shard of energy in his body to reach back toward life. So far, he's made it back alive every time.

The four rules to his formula—commitment, crash, submission, and escape—provide a blueprint for every failure in his life, he says. It has allowed him to survive the unsurvivable and has given him a guiding philosophy for every venture he gets into outside of the water. "It's the attitude I go into before I start a new business, or a new training regimen," he says. Together they form a wedge that he crams between his consciousness and the autopilot of his own nervous system. It's the division between his own natural limits and the control he can express over the world around him. Hamilton's philosophy mixes spirituality with exercise and environmental conditioning. An outsider might call it holism.

Sometimes when I hear this sort of all-encompassing explanation for a universal order I can't help telegraphing my skepticism. Is one of the world's greatest surfers also a purveyor of Californian flim-flam and fad exercises? An easier explanation for his success, after all, could be that he has access to the very best training facilities on the West Coast and that he's just been incredibly lucky to survive every death-defying feat. However, scientific explanations and claims can be equally mysterious. And there's no better illustration of how the holistic world and scientific world have collided than the recent research gold rush into brown fat.

7

SWATTING MOSQUITOES WITH HAND GRENADES

IN THE PAST hundred years the rise of easy calories, climate control, and processed foods have created a global obesity epidemic and an exponential rise in diabetes. In the United States alone, diabetes accounts for $245 billion of annual medical spending. Almost 35 percent of Americans are obese, and the rest of the world's citizenry isn't much better: 39 percent of the rest of the planet's population is medically overweight.[1] We spend fortunes on crash diets, and, when they don't work, billions more on kidney transplants and dialysis. We've tried just about everything to bring the crisis to a halt: from inventing drugs that limit the amount of fat the body can absorb to amphetamines that overclock a person's metabolism to even intentionally infecting ourselves with tapeworms. Yet none of this has slowed the steady expansion of our collective waistline.

---

1   Doctors define "overweight" as a body mass index, or BMI, of 25 or greater. A BMI of 30 or greater is considered obese.

Then, in 2011, scientists at Harvard discovered that brown fat wasn't just a vestigial tissue locked away in our evolutionary past, but was something that just about every human has the potential to build and use. It was a ray of hope amidst a bleak and unrelenting epidemic. Within a year, the National Institutes of Health issued a new research directive to help fund basic research into the underlying mechanisms of brown adipose tissue (BAT) with a simple long-term goal: to create a drug that maximized all the properties of brown fat without requiring someone to spend their time being cold. The first pharmaceutical company to develop a drug that could turn on BAT's fat-sucking properties without changing the sedentary American lifestyle would not only stand to do a great deal of public good, but would make billions of dollars in the process. The race for a miracle drug was on.

The first task was to understand exactly how BAT works. Since it was only recently discovered in humans, many of the basic mechanics were still a little hazy. But the broad strokes of the process seemed fairly simple. Once nerve cells in the skin detect a cold environment, they relay a chemical signal through a series of neurons leading to the autonomic centers in the brain with a message to get ready to start heating the body. The brain then has a few potential options to deal with this information. It can simply ignore the signal and calculate that the body's stores of white fat provide sufficient insulation, and because the brain perceives no real threat it does nothing. Or the brain can trigger various muscles to start shivering and mechanically heat the body. Or it can ramp up the metabolism and tell brown fat to start sucking white fat out of storage and burn it for heat energy. Or it can decide to do a mix of all three.

Since sensory nerves relay the signal to the brain, there is also a conscious component. When you feel cold you perceive it as a slight chill, which might lead your conscious mind to override any autonomic decision-making and instead start some sort of exercise, turn up the thermostat, or put on a few additional layers of clothing. No matter what happens, at the most minute level all the signals and sensations break down into a cascading series of hormone and neurotransmitter releases that stretch from the nerves in the skin to the brain, which then triggers one heating strategy or another. In theory, a drug could short-circuit the brown fat activation process by mimicking the specific neurotransmitter that the brain sends out along this chemical pathway to activate BAT. A person would simply take a pill or an injection—skipping any discomfort that accompanies being cold—and start burning up ordinary white fat.

The tricky part is identifying the right series of neurotransmitters and chemical pathways. One company in Boston aims to create a special line of stem cells that will turn white fat cells brown, thereby giving patients enough BAT in their systems to help kick-start passive thermogenesis. A company in Houston believes they have a promising candidate for a drug that will mimic the correct neurotransmitter to turn on BAT. Meanwhile researchers in the San Francisco Bay area have discovered that certain environmental conditions will make the body turn white fat "beige" (think khaki) by a not-yet-understood process.

Some recent successes are particularly startling. Kevin Phillips, PhD, assistant professor of molecular and cellular biology at Baylor College of Medicine, started thinking about how rodents have a special ability to transform their own stores of white fat into brown fat while preparing for hibernation. Phillips and his team knew that an overactive thyroid gland makes it difficult for people to tolerate heat, while an underactive one makes them especially sensitive to the cold and wondered if the key to turning on brown fat lay in the thyroid itself. He created a synthetic thyroid hormone—called GC-1—that would stimulate production of a protein in brown fat called UCP1, which is the critical component that allows BAT to dismantle white fat cells and turn them into heat. Phillips located mice that had been genetically programmed to be obese (apparently made-to-order mice exhibiting just about any condition are easy to come by in the halls of scientific research), and injected half of them with his experimental compound. Over the course of the next 20 days the mice that received GC-1 lost almost half of their white fat while the control group actually gained weight. To illustrate the change, he published photos of the mice in the November 2015 issue of *Cell Reports* showing

These two genetically identical mice were kept in a cold room for several days. The mouse on the left was given a drug named GC-1 that stimulated brown fat production and activation. The mouse on the right is the control group. The results indicate that brown fat can be a key driver in weight loss. (Photo by Kevin Phillips)

one sad looking, obese fur-ball with its stomach flattening out across a white table beside a trim counterpart almost half its size.

Along with the startling weight loss, the mice with active BAT also had an innate ability to resist the cold where the control group did not. In a cold tolerance test, Phillips put both groups of mice into a refrigerated cage at 39.2 degrees Fahrenheit and measured their body temperature. Over the course of 8 hours the obese control group's core temperature dwindled into fatal hypothermia. The mice that were dosed with GC-1 continued onward for 15 straight hours, apparently no worse for the wear. Phillips published his findings in *Cell Reports* and then to great fanfare in the mainstream media for uncovering what might be a way to bottle brown fat's lightning and give people a simple pharmaceutical miracle pill. He plans to conduct a follow-up study on nonhuman primates, and if all goes well with those, on to humans.

While one day a drug could well hack the body to create stores of brown fat, there is reason to pause before declaring this a solution to the ailments of the modern world. Any sort of drug—even ones very carefully designed by the best researchers on the planet—are clunky, cell-specific delivery mechanisms at best. The communication chains in the nervous system work a little bit like pieces of a puzzle in that one neuron activates its neighboring neuron by excreting a specific chemical directly into the space between them. This space is called a synapse, and this

Wim Hof meditates on a glacier somewhere north of the Arctic Circle.

Wim Hof's skin burns red while climbing to the summit of Mount Snezka.

photo by Scott Carney

Wim Hof emerges from an icy lake.

photo by Henny Boogert

Wim Hof, Janis Kuze, Vladamir Stojakovic, me (Scott Carney), and Andrew Lescelius meditate below a freezing waterfall outside the Innerfire training center in Poland.

That's me on snowshoes in the mountains of Idaho.

At the summit of Mount Snezka after an arduous shirtless climb up the slope.

Hof, me, and other trainees sit on the banks of the river until the snow melts around us in Poland.

photos by Jeremy Liebman

I don a mask at the CU Sports Medicine and Performance Center to test my carbohydrate- and fat-burning abilities.

Laird Hamilton lifting dumbbells underwater at his pool in the Malibu Hills.

Laird Hamilton and his wife,
Gabrielle Reece, train beneath
the surface of their pool.

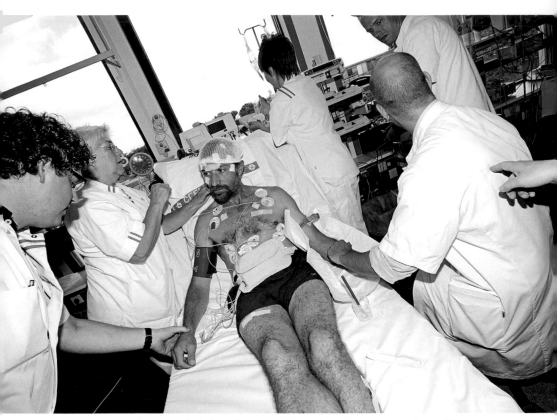

Scientists Matthijs Kox (far left) and Peter Pickkers (far right) prepare to inject Wim Hof (center) with endotoxin at Radboud University in the Netherlands in order to investigate his claim that he can consciously suppress his immune system.

Hans Spaans has suffered from Parkinson's disease for 12 years and claims that the Wim Hof method has allowed him to drastically reduce his dependence on pharmaceuticals. He sits here in a squat before his morning calisthenics routine.

Henk van den Bergh suffered from crippling rheumatoid arthritis that had left him immobile at home. His condition has drastically improved since adopting daily ice baths and a rigorous breathing routine.

Brian MacKenzie slowly lets air escape
his lungs underwater at his training
center in Orange County.

photo by Chris DeLorenzo

The November Project takes over Harvard Stadium in October 2015.

Bojan Mandaric, cofounder of the November Project, rests after an intense fall workout.

Military researcher John Castellani wears a balaclava as he tests his manual dexterity on a Purdue Pegboard at USARIEM in Natick, Massachusetts.

Ed Gamester—dressed as "El Nordico"—delivers a high-flying kick to Madhi Malik two nights before Tough Guy 2016.

Mr. Mouse, aka Billy Wilson, tests the electrified wires before Tough Guy 2016.

photo by James Appleton

Me climbing over a
cargo net at Tough Guy.

photos by James Appleton

Filmmaker
Scott Keneally
and I pause for
a picture in
the middle of
the obstacle
course race.

Me, Wim Hof, and Dennis Bernaerts at Gilman's Point on Mount Kilimanjaro.

process happens millions of times in sequence with exactly the right locks and keys to activate a predetermined biological process. Drugs try to hack this chain by flooding the entire body with synthetic hormones that mimic the correct chemical shape, but they do so without entering the chain in the correct sequence. Unfortunately, the body only has a finite number of receptors and shapes available to it, and the same protein or neurotransmitter for, say, activating brown fat could also be critical for a wildly different process in another part of the body. Normally the dearth of chemical shapes isn't a problem, because the pathways are kept separate and individually activated. However, when a drug floods the bloodstream with a non-native molecule, there is no telling where it might end up. Sure, it can find the correct site on the BAT pathway, but it might also bind to other sites and wreak havoc in the process. In other words, every drug has a potential for side effects. In a way, the drug approach is a little like trying to swat a mosquito with a hand grenade. Sure, a grenade will kill a mosquito, but who knows what else might get caught in the explosion.

Over the course of experimentation and clinical trials researchers attempt to minimize and rule out potential side effects. But anyone who has read the medical literature on just about any drug on the market knows that there's always potential for unexpected reactions.

But there's another reason that the pharmaceutical approach to activating BAT might not work as well as expected, and it has to do with a fundamental rift between most laboratory research and the complexity of human experience. Living things perceive the world through sensation. We feel pressure, cold, pain, and warmth. We can see and hear things, and all these senses help our minds evaluate the world around us. However, from a medical perspective there's no way to specifically identify these sensations. In the same way that a microscope or probe can't touch a thought or poke at pain, no scientific method yet developed can bridge the gap between the mind and the body.

As Galen Strawson, author and professor of philosophy at the University of Texas, astutely pointed out in his May 2016 article for the

New York Times that German philosopher Gottfried Wilhelm Leibniz described this disconnect quite vividly all the way back in 1714. Consciousness is "inexplicable on mechanical principles, i.e., by shapes and movements," Leibniz wrote. "If we imagine a machine whose structure makes it think, sense, and be conscious, we can conceive of it being enlarged in such a way that we can go inside it like a mill. Suppose we do: visiting its insides, we will never find anything but parts pushing each other—never anything that could explain a conscious state."

The scientific method equates sensations and thoughts to chemical signals and physical processes. This sort of reductionism makes it almost impossible to assign meaning to experiences. Moreover, by focusing so closely on single isolated neurological pathways it's possible that this approach misses the confluence of bodily interactions with the environment that, in turn, create the overall biological picture. As Phillips's investigation noticed, the thyroid certainly plays a role in how humans interact with ambient temperatures. Cold can trigger brown fat production, but it can also initiate a number of other responses throughout the body depending on which biological strategy seems appropriate in the moment. Put another way, even if there are no side effects from a drug that ramps up brown fat, there's no reason to think that brown fat production alone is the only benefit of cold exposure.

Wouter van Marken Lichtenbelt, a professor of energetics and health at Maastricht University in the Netherlands, has tried to turn a scientific lens on the questions of how humans adapt to different environments. I've been trying to connect with him for several months but have had very little luck finding a time when our two schedules lined up. When I do manage to eventually locate him through a scratchy cell phone connection, he is riding a bicycle on his daily commute between his home and his laboratory. I can hear the brakes squeal over the connection as he comes to a stop somewhere beneath a highway overpass. "It's a cave, really," he comments as he peers into its dark recesses. "It appears that I'm in the company of bats now." It is a fitting spot. It is winter in Holland and the bats around him are in deep hibernation. In order to preserve their energy through the winter months, bats and many other small animals don't shiver but instead warm their bodies by burning up white fat. "I've been interested in non-shivering thermogenesis for almost ten years, and have been looking at ways to bottle metabolic lightning at least that long," van Marken Licktenbelt says.

His studies aim to understand the way that the environment alters the body. While pharmaceuticals might be able to target specific bodily processes and tweak them to treat a specific acute ailment, van Marken Licktenbelt sees the twin epidemics of diabetes and obesity through a thermodynamic lens. Both conditions come from there being too much energy available for the body. When there's too much energy in a system, then that system starts to malfunction. It's a matter of physics. And there are really only three ways to fix it: You can reduce the amount of energy coming into the body, decrease the body's efficiency of absorbing energy, or increase the rate that the body makes use of it. Most diets follow the first line of attack. To lose weight, you can simply significantly decrease your food intake. Working on the input side of the equation will result in some short-term success, but more often than not people struggle to restrict their calories and end up regaining their lost pounds over time. Several drugs follow the second line of attack by changing the efficiency with which the body actually processes food. In the 1920s, magazines sold tapeworms for people to ingest and host. With the help of the worm, people could eat whatever they wanted and still lose weight. But parasites are not the most friendly companions, and this treatment often led to severe anemia and malnutrition. A more recent iteration of the same idea is a drug called Xenical. It sets an artificial limit to the amount of fat that a person can efficiently metabolize. It's what Bill Clinton supposedly used to slim down after a presidential diet of Big Macs. While this drug is effective at curbing the amount of calories a person can absorb, there is a nasty side effect: All the excess energy transforms immediately into an oily waste. In practical terms this means that just a few extra calories of fat sends patients careening in clench-legged sprints toward the nearest bathroom. The final option is, of course, ramping up the metabolism, and van Marken Lichtenbelt believes that the best way to do this is through a combination of exercise and cold acclimation.

Diabetes is essentially a condition wherein the body cannot deal with the amount of sugar in its bloodstream. As sugar levels go up, healthy people produce more insulin to control the influx. But people suffering from type 2 diabetes grow resistant to insulin so that sugar levels rise. This excess of sugar wreaks havoc on their insides: limbs swell, cankerous lesions form, the extremities become numb, and ultimately the patient's kidneys are destroyed. Doctors have nicknamed the disease "sugar cancer." In 2015 van Marken Lichtenbelt found eight overweight men in their late fifties suffering from type 2 diabetes and decided to test them to see how short-term cold exposure would alter their bodies. The men dressed in shorts and sat in a chilly, 57 degree fahrenheit room, the temperature just above the point where they would begin to shiver uncontrollably. They spent 6 hours a day in the cold for 10 straight days as the doctor monitored their insulin production and blood levels. At the end of the study he put the men into a PET/CT scan to measure their levels of BAT. As he suspected, at the end of the study the men showed enormous improvements in how their bodies metabolized sugar—the cold allowed the men to clear sugar from their blood 43 percent more efficiently than when they started.

In other words, in just under 2 weeks cold exposure almost reversed the symptoms of diabetes. As one of the pioneering researchers on BAT, van Marken Lichtenbelt predicted that the metabolic changes would come with a corresponding increase in their BAT levels, but when results from the PET/CT came back it turned out that most of men had the same levels of BAT as when they entered the study. Cold exposure had changed the men's underlying condition, but the lightning that van Marken Lichtenbelt was trying to bottle must have come from somewhere else. The only thing that he could conclude was that the men's bodies had discovered a different way to adapt to the cold conditions than what he had anticipated. So while the study had enormous implications for how to treat diabetes, the underlying mechanisms were still clouded in mystery.

The miraculous properties of BAT are just one of many strategies that the body can rely on to accomplish the same tasks. An extremely overweight person with a surplus layer of insulating white fat doesn't necessarily need to ramp up their metabolism to fight against the cold. The layer of fat alone suffices to protect against moderate exposure.[2] People who are already in shape, however, might respond much more quickly to cold temperatures, forcing their bodies to generate BAT in order to stay alive. After a series of biopsies to investigate what other possible mechanisms might have helped his patients, van Marken Lichtenbelt discovered that ordinary muscles can also transform under the right conditions and hyperstimulate other cellular mitochondria to accomplish the same results as active BAT. The secrets of the human body are complex and highly dependent on the individual and the method of exposure even if the results are the same. However, more than anything else, this study shows that while a drug might activate one specific pathway to generate heat energy to burn up excess calories, administering that drug might not be the optimal solution. Instead of stimulating the body with a foreign substance, it is probably more efficient to use the environment to trigger the correct and predictable response.

To be fair, van Marken Lichtenbelt had some inkling that even though brown fat was potentially important, it was not necessarily the only key to large-scale metabolic change. As seems to be the case with just about every biologist in the Netherlands, early on in his career he had a chance to study Wim Hof's curious physiology.

In what might just be the most profoundly lucky occurrences of happenstance for a researcher investigating Hof's method, it turns out that Hof has a genetically identical twin brother, André. Previous examinations of Wim Hof's physiology had shown that the Iceman himself possesses much more BAT than other people his age. He has about the same

---

2   Presumably, the mice that had their BAT activated through the chemical in Phillips's laboratory spent a lot longer in the cold than what could be dealt with solely through insulation.

amount that a typical 20-year-old does. While Wim Hof is incredibly active, his twin brother isn't. As a truck driver who spends most of his time inside a heated cab, André lives a mostly sedentary lifestyle. Though they look the same, van Marken Lichtenbelt figured that Wim's near-constant exposure to the cold would result in his having a much greater amount of BAT than his twin would. But when the professor put both Wim and André into a PET/CT scan, he found that they— remarkably—had very similar (high) levels of BAT. Even so, despite their similar biology, Wim's ability to withstand the cold is undoubtedly stronger than his brother's. While BAT likely plays a part in his metabolic robustness, defining the exact underlying mechanisms proves to be a deceptively difficult task.

All this, and still the most striking discoveries on the Wim Hof Method don't have anything to do with his metabolism. In 2011, Hof met with Dutch immunologists Peter Pickkers and Matthijs Kox at Radboud University after making the outrageous claim that he could consciously subdue or ramp up his immune system at will. It was a claim that was, by definition, impossible. The prevailing medical logic at the time held that there was a firewall between the autonomic and somatic nervous systems. The immune system wasn't even supposed to be connected to the brain at all. Nonetheless, Kox and Pickkers were curious, and if anyone could test Hof's claims it was them. Until this point much of Pickker's career had been devoted to developing tests that evaluated the effectiveness of immunosuppressive drugs. While turning off the immune system is not usually a great idea, in some cases—such as when someone gets a kidney transplant and their body might reject the donor organ, or in the face of an aggressive autoimmune disease—there is no other way for a person to survive. In 2011 Kox was Pickker's graduate student looking to finish his PhD and distinguish himself in the medical community. The test that they devised aimed to trick a person's immune system into believing that it was infected with a deadly strain of *E. coli*. Under normal circumstances, once the immune system detects *E. coli* it starts to produce antibodies and mounts an aggressive fever response to stop the infection before it spreads. People whose immune systems are already compromised, such as by a drug or sickness, continue on as if nothing happened. So the test he devised for Hof was simple: He'd inject him with the dead bacteria and see how he responded.

As I mentioned in Chapter 1, when the team injected the solution into Hof's bloodstream he showed almost no reaction at all—a result that astounded the scientists and helped earn Kox an award for his PhD dissertation on anti-inflammatory pathways. If the results held up to scrutiny, there would be enormous implications for anyone suffering from an autoimmune disease. However, the scientific community was far from willing to admit to a medical breakthrough or start rewriting medical textbooks. The first and most prevalent criticism of the Pickkers and Kox study was that perhaps Hof was a genetic anomaly. Sure, maybe Hof could affect his immune system, but he was probably just an exception to the rule—some sort of freak of nature, not a miracle worker. So in 2012 Pickkers and Kox designed a second experiment. This time they wouldn't test Hof. Instead they would tell him to teach other people his technique and see if those students achieved similar results.

At first blush you would think that locating 30 people to volunteer for a bacterial injection that makes 99 percent of people feel terrible would be an uphill battle. But when the Dutch researchers announced the impending study at the university campus and told students that they would have a chance to study with Wim Hof, they fielded more applications than they knew how to handle. The study split volunteers into two groups. Twelve people in the control group would go about their normal lives in Holland, while the second group of 18 would travel to Poland to study Hof's techniques of ice baths and breathing for 10 days. Predictably, no one wanted to be in the control group, so Hof volunteered to teach his methods to the people in the control group (those who wouldn't go to Poland) after the test was over.

One week after I left Poland after summiting Mount Snezka, three instructors flew to the farmhouse and Hof taught the active group three basic techniques: cold exposure in the snow, focused third-eye meditation, and sequential muscle retention after hyperventilation. They climbed up the same mountain that I did and baked in the same sauna. When they returned from their trip the volunteers continued to practice on their own for 5 days before showing up to Pickkers and Kox's lab for supervised injections. The results were astonishing.

Even after such a short training program the active group showed positive levels of epinephrine as well as an increased amount of anti-inflammatory molecules in their blood. They had fewer fever-like symptoms than the control group experienced, and their cortisol levels returned to normal much quicker. To quote the subsequent journal article from the *Proceedings of the National Academy of Sciences:* "Hitherto, both the autonomic nervous system and innate immune system were regarded as systems that cannot be voluntarily influenced. The present study demonstrates that, through practicing techniques learned in a short-term training program, the sympathetic nervous systems and immune system can indeed be voluntarily influenced." This short declarative statement forced the scientific community to completely reevaluate their understanding of the immune system. The article earned a mention on the journal *Nature*'s website and caught fire across the internet, lending scientific credibility to Wim Hof's program. If the finding continues to hold up to scientific scrutiny, then it would seem to have potential implications for a huge variety of illnesses—from autoimmune conditions to diabetes to bacterial infections to food allergies to, well, anything. If not an actual cure for any of these, environmental stimulation adds an important dynamic to the overall picture for treating human illness.

A year later a related but unaffiliated anatomical discovery helped shed light on how conscious control of the immune system might work. For the most part, the immune system has a standard operating procedure for uncovering and destroying foreign invaders. Once it locates a virus, tumor, bacteria, or, in the past 60 years, transplanted organs, it sends out white blood cells to eat the pathogens or explode the sick cells. The process is essentially the same in every part of the body except for the brain. If an infection breaches the blood-brain barrier (think meningitis), the body has no defense against invading bacteria. Presumably evolution was to blame: the brain was simply too fragile to allow the immune system to meddle inside of it. More importantly, there appeared to be a firewall between them. Every anatomical textbook posited that the immune system (or, more specifically, the lymphatic system) and the brain were completely separate and had no physical connections.

However, in 2005, Jonathan Kipnis, a neuroscientist at the University of Virginia, noticed that mice with severely degraded immune systems also suffered severe cognitive impairment. He discovered that mice with low T-cell counts forgot to run through mazes that they used to traverse with ease. He also learned that if he boosted their immune systems, their mental abilities improved. This led him to hypothesize that there must be a connection between the immune system and overall brain function, and that perhaps compromised health could explain the cognitive decline of people with HIV or dementia. But the findings puzzled Kipnis because they contradicted the notion that the brain and immune system were separate. Even though he could show that the brain wasn't working well with a compromised immune system, there was no anatomical reason for that to be the case. So he did what any good biologists would do in his position: He began dissecting mice.

In 2014 Kipnis started to look at the vertebrae along mouse and human spines for any sort of gateway where T-cells might come into contact with the central nervous system. There he discovered small, fluid-filled lymph nodes that shouldn't have been there. They didn't appear in any medical literature, and he realized instantly that he'd made a major discovery. New anatomical discoveries are incredibly rare in the modern age—doctors have been dissecting humans and mapping internal anatomy since at least the time of Leonardo da Vinci—but these sacs connected directly into the spinal cord and served to drain lymph from the central nervous system. In 2015 the most prestigious scientific journal in the world, *Nature,* published his findings with a call to rewrite medical literature, stating: "The presence of a functional and classical lymphatic system in the central nervous system suggests that current dogmas regarding brain tolerance and the immune privilege of the brain should be revisited." For the first time, it was clear that the brain and the body were not so different after all.

Without a barrier between the two, even the most reticent researchers have to admit that it is at least possible that people might be able to consciously affect their immune systems. Indeed, as Kipnis initially suspected, medical literature is full of examples of how schizophrenia and even something as mild as psychological stress can make people more vulnerable to immune problems. Why couldn't that communication travel in the other direction? A person in good spirits and environmental balance should be able to get an immune benefit.

The strangest thing about these scientific discoveries is that outside the strict confines of ivory tower journals, the thought wasn't really a revolution. It's something that we knew all along.

8

IT'S RAINING INSIDE

NO ONE NOTICES when the first drops of water begin to form on the ceiling. I am standing in the upstairs shower warming in a cascade of hot water after my morning dip in Hof's ice-cold pool. The house is a brand new acquisition to the Iceman business and there are still a few kinks to work out. One of which is that the drain in the middle of the bathroom floor tends to get clogged with debris. As I stand there oblivious to the potential problem, water pools in the center of the room until it overflows a marble threshold tile and then spills out into the hallway. From there it follows the will of gravity, slipping between the cracks in the wood, and seeping into the space between the subfloor and downstairs ceiling. The soapy water collects in the cavity until it saturates the materials below.

When Hof bursts into the bathroom, my first thought is that he somehow knew I'd dared heat myself up in a shower, and that he is going to scold me until I turn the faucet to a temperature that will shock my nervous system into submission. Instead he yells at me in a mixture of Dutch and English before pushing me aside and turning off the knob. It is only my second day in Holland and I've already flooded his brand new house. I wrap myself in a towel and head down the staircase to survey the damage myself.

Hundreds of droplets hang precariously from the ceiling. Each slowly elongates until it grows too heavy and comes splashing down into the middle of his living room. They pepper the sofa, his guitar, and his television. Hof's dog, Zina, barks happily at the impromptu rainstorm. There are three of us in the room: me, Hof, and a redheaded American acolyte named Anton Nicola who has just finished a recent winter training session in Poland. Together we push furniture out of the way and scrounge rags and towels from the bare closets to sop up the deluge. Minutes pass and if anything the storm just grows stronger. A knot forms in my stomach when I realized that the ceiling might collapse. It is all my fault, but when I look over at Hof he is standing in the middle of the torrent with his hands raised over his head. A broad smile beams across his face when he starts shouting.

"It's a miracle! Can you believe it's raining inside my house? What luck!" Hof dances amid the droplets, letting the water fall where it might like some sort of deranged rain god. It's not the reaction I expect or that I would mete out to my own guest in the same situation. But for Hof, the destruction of his home marks an inexplicable moment of joy.

Jump back to a few days earlier. It is Black Friday—the day after Thanksgiving—when all of America participates in a communal shopping orgy as they open their wallets, unable to resist the lure of discounted consumer goods. For the previous month I've kept the thermostat at my house at a steady 62 degrees, a full 10 degrees lower than the year before, when my wife used to chide me that I was perfectly happy to hole up in my office with the space heater on. It has been almost 4 years since I'd first met Hof. Although I have become better at my practice since this summer began, I must confess that my training program since first meeting Hof in Poland could politely be termed irregular. Some mornings I wake up and fall effortlessly into breathless breathing routines with a cold shower finish, while other times it is so much easier to wipe the sleep from my eyes, pour a cup of coffee, and move on to other things.

Once the habit is broken and the days march forward it can sometimes take a week before I take it up again. Who am I kidding? It might be a month. Perhaps that's the reason I book a ticket to see Hof. I need a shot of motivation.

So I sit on the tarmac at the Denver airport looking out the window at a sky that can't decide whether to drop snowflakes or icy sleet. The plane engines hum as a fleet of de-icing machines descend on the aircraft like droids from a dystopian future. The vehicles are little more than tankards of antifreeze with thin hydraulic arms that hold workmen in small, elevated glass cabins. The contraptions spray down the wings with chemical cannons. Four such machines coat the plane until it is so slick that the wings reflect the glare of the yellow airport lights. Unlucky snowflakes that make contact with the wings vanish in puffs. Humanity's fight against the elements always has new fronts. Without the chemical spew, ice might undermine the delicate aerodynamics of the aircraft and send us careening into the Atlantic. And so I begin to wonder how my own fight against the ice will go. Nine hours and a connection in London later, Enahm Hof, Wim's son and business manager, meets me at the Schiphol airport wearing a wide grin.

Enahm leads me to a small Audi sports car, and as we take off down the highway he explains how he's transformed the chaos of his father's teachings into a coherent—and profitable—business called Innerfire. When we'd first met in 2012, Wim had not yet achieved international fame. Since then a wildly popular documentary on the HBO series *Vice*, countless podcasts, news articles, and maybe my own story in *Playboy* had spread the message across the world. Would-be followers wanted to learn how to perform similar feats in the ice, and a growing scientific literature was lending credence to his claims that the method could hijack the autonomic nervous system. A key driver in this sudden interest was that learning the technique isn't very difficult to learn. All it takes is a little bit of mental fortitude and strong lung muscles. But what alchemy allows ice baths, mental focus, and deep breathing to crack into the autoimmune system? The science is sometimes confusing, even contradictory. Despite my experiences with the Wim Hof Method, it still seems too good to be true, and the skeptic in me still wonders if perhaps it is. Maybe there is a secret, deeper technique that I haven't picked up yet. There *has* to be more. It is one thing to meet already-impressive athletes who claim that the method pushes them to a new level, and it is quite another to find that level yourself. A measly ski hill in Poland isn't enough. I want my own Everest, some impossible feat that I can overcome for myself and afterward look out at the world a little stronger than I have any right to.

In January Hof plans to summit the tallest mountain in Africa—Mount Kilimanjaro, a soaring 19,341 feet above sea level—in just two shirtless days. The trek won't be all that difficult from a mountaineering perspective, but doing it Hof's way is a serious test of human mettle. Kilimanjaro isn't formidable because of rockslides or long rope ascents on bare rock walls, but for its high incidence of altitude sickness in expeditions. As a hiker trudges upward, the air grows ever thinner and the atmospheric pressure ever lower. Eventually the air gets so sparse that oxygen can't circulate efficiently throughout the body. When the dizziness progresses to full-blown acute mountain sickness—AMS for short—the heart starts to beat irregularly and the lungs can fill with fluid until it becomes impossible to breathe. Five to 10 people die every year on Kilimanjaro ascents. To stave off this condition mountaineers slowly acclimate on their way to the top, ascending in easy stages so that their bodies can produce more red blood cells and get used to the thinning air. Alternately they dose themselves with a drug called Diamox, which helps them adapt to the altitude. A typical Kilimanjaro ascent takes a minimum of 5 days, plus another 3 or 4 to get back down. Even going at that pace, only about 45 percent of people who attempt the summit make it. Hof claims that with focus and concentrated breathing he can take 20 people to the summit at a pace that most mountaineers would call reckless. But then again it sounds exactly like the sort of challenge I need to prove that the training has actually changed me.

But my plans hit a snag. For reasons I do not entirely understand, a few months earlier Enahm put the kibosh on my request to accompany Wim to the top. The trip is mysteriously full every time I ask, and there is never more space on its roster. I have a feeling there is a deeper problem, and one of my goals in coming to Holland is to secure a space on the expedition.

The journey from Amsterdam to the training center takes almost an hour. Enahm pilots his sports car through a densely packed highway of economy cars, swerving through traffic, and expertly slowing down for automatic cameras that are set up to catch speeders like him. The center is located just outside a tiny village called Stroe and has been part of the organization for only a few weeks. It's just one sign of the exponential growth of all things Wim Hof in the past 4 years. Early in his career Hof suffered from a series of bad business deals with a string of managers who pushed him into shooting stunts that garnered media headlines but which made people think of him more as a circus act than a metabolic genius. As he climbed most of the way up Mount Everest and ran across the Arctic barefoot, various managers pocketed the lion's share of the proceeds. Enahm took over the operations in 2010, founding Innerfire and urging his father to do fewer stunts and instead lend his body for scientific research. The move made Hof's affairs a family business, and under Enahm's direction the company produced a 10-week online video course that spread virally in online forums and through tens of thousands of Facebook posts. Together they created a standardized teacher-training program to certify people as Wim Hof Method instructors who could spread the message without needing direct supervision from Hof himself. In the next few days 100 of them would take a midterm exam in Stroe to see if they'd qualify to take the same Poland course I completed several years earlier.

Now the Iceman is a household name in the Netherlands. His workshops sell out within minutes and lecture venues easily fill with 600 souls. Every month about a thousand people sign up to take his 10-week online course. It is hard to imagine how it would have grown as large as it has without Enahm taking the reins. Under his direction the business expanded operations from the Netherlands, Poland, and Spain to an intercontinental audience across the Atlantic. For a journalist looking to tell Wim's story, however, Enahm can be something of a roadblock. Part of Enahm's vision is to keep the message about the method on point—focused on science and less so on the aura of chaos that seems to follow everything that Wim touches. Because I am someone who has seen Wim emerge from the shadows, Enahm might think of me as a bit of a liability. Or, perhaps Enahm is simply worried that I might not make it to the top, something that certainly wouldn't do Wim's image any good. So with this in mind I keep the conversation light. The sun is just going down as we pull into the gravel driveway.

The new training center is a converted metalworking shop that hasn't quite made the transition to new-age ashram. The two-story villa still has paintings of the previous owner's children hanging on the wall. Its hallway opens into a barren kitchen with heated floors and a dog-hair encrusted sofa. A pervasive smell of mildew emanates from every crack in the walls, and only a couch, flat screen TV, and a broad, leafy houseplant make up the furnishings in the living room. "This is all going to change," Enahm says of the rough accommodations. "We have a meeting with an architect in a few days." What I am seeing is at best a work-in-progress. It isn't at all clear where I am going to sleep, and when I ask Enahm he just shrugs and waves me upstairs.

Of the four rooms on the second floor there is only one piece of furniture: a bed with a broken spring in the middle. At its foot is the sum total of Wim's personal wardrobe: a white plastic trash bag full of dirty clothes, a hopelessly wrinkled white sports coat, an orange swimsuit, and a few towels. The pile sits discarded on a stained brown carpet. I stare at the collection for a minute. Part of me despairs what a week in such Spartan quarters will feel like, while the other part marvels at the disjunction between the worldwide fame and riches, associated with the training empire, and the way Hof chooses to live. Wim is many things, but he is clearly not motivated by material trappings.

Enahm calls up from downstairs. "I think my dad usually sleeps on the couch," he says. "You can probably take his bed." So I put my rolling suitcase on the floor. Enahm leaves and an hour or so later, around midnight, Wim pulls up outside in a large, yellow Ford van. He comes inside wearing a T-shirt, nurse scrubs, and a black suit jacket. Next to him is Anton, the tall, redheaded American whose mop of hair resembles that of the comedian Carrot Top. Anton is traveling around Europe after a stint learning Wim's method and is trading accommodation for some work around the house. Considering the bare bones accommodations, I wonder if we'll be in competition for the only bed.

"Scott Carney, you motherfucker!" Wim yells into the small room with a jovial smile on his face. It is meant to be a compliment of sorts. The curse words are the newest tools in his English arsenal and have been reprinted a thousand times on a T-shirt with the company logo interposed with the slogan "Breathe, Motherfucker!" Wim has no time for pretense and sometimes jokes with his instructors that the people who completed his courses earned their M.F. degree when they were done with his training. The insult, which might make me angry in another context, instead provokes a cracked smile on my face. You know what? It has been far too long since I've seen this motherfucker.

For the last few hours he's been trying to set a new Guinness world record for getting the largest number of people to ever walk on an ice rink barefoot. The event took place in a small village on the other side of Holland. By the time it was over the crowd's communal body heat melted the top layer into slush. Hof is still energized by the sight of the drag queens, the school groups, and the endless streams of well-wishers who have been drawn into his orbit.

Although it is late, Hof has the vitality of a man a quarter his age. We sit together on the couch and chat for several hours. In the last few years Hof went from being an unknown to a world-famous guru, and he says he is struggling to understand what it means to be an icon as well as the figurehead on the prow of a several-million-dollar-a-year business. The constant lecturing, workshops, and media appearances are not what he is built for. He'd much rather be a madman than a prophet. But with great power comes great responsibility, so Hof has to zero in on the kind of highly refined message that can captivate the largest possible audience. He tries a slogan on me; it's just three words.

"Healthy, happy, strong," he says with pride. "I just want to keep it simple."

These, he says, are the three ingredients to a good life. Forget, for a minute, the science behind breathing and the cold, the barbaric discomfort of ice water, or how many pushups a person might do without breathing. They're all just trappings of a larger desire to live well. From there we descend into a more detailed discussion of what it means to live a good life. Hof starts to speak quicker, like he is agitated. He begins to wave his arms in the air as if he is keeping a small insect at bay. "Society is sick," he says.

"Have you ever seen a rabbit go to a pharmacy, a hospital, or a mental asylum?" he asks rhetorically. "They don't look for medicine, they heal themselves or die. Humans aren't so simple; they've let technology get in the way of who they really are." It's an idea that I've thought a lot about, and one that doesn't always sit comfortably. Yes the modern world has its drawbacks, but nature can also be brutal. So I interrupt the budding diatribe.

"But rabbits get eaten by wolves," I say.

Hof doesn't skip a beat at my interjection. "Yes, they know fight and flight. The wolf chases them and they die. But everything dies one day. It is just that in our case we aren't eaten by wolves. Instead, without predators, we're being eaten by cancer, by diabetes, and our own immune systems. There's no wolf to run from, so our bodies eat themselves."

This is, of course, at the core of Wim's philosophy. Without something to fight against, the body will fight itself. They are words to muse over for the next week. Before I shuffle off to bed Wim offers to let me tag along for a morning excursion with the Dutch Olympic sailing team as they make their preparations for the 2016 Summer Games. He is going to lecture them on what it means to be strong and how to stay at the top of their game in any weather. Eight hours later I take my inaugural shower and transform his house into a rainstorm.

Wim's dance only ends when the rain from his ceiling slows to a trickle. Fortunately it doesn't actually collapse. In the joy of the moment he forgets to check the time and realizes that we need to be on the other side of Holland in less than an hour. I'll be sharing a seat with his dog in the cab of his bright yellow van. A minute or two later we are on the highway trying to transmute an hour into 40 minutes by stepping on the gas. "Nothing to worry about," he smiles as his blue eyes focus on the road. "They're gonna learn patience today, too."

The drive gives us time to talk and for me to slip the idea that perhaps it would be fun if I joined the Kilimanjaro expedition. The group—which has grown to 26 people—all agreed to practice the method for half a year and meet up every few weeks to breathe and then jump into one of Amsterdam's canals or shallow lakes. Sure, I've missed some of the camaraderie, but I live at a higher altitude and have been training on and off in the method since we summited Snezka. How much harder could this mountain be?

Hof doesn't skip a beat responding. "Of course," he says with his usual enthusiasm. "Not only should you come, you *must* come." He starts punching buttons on his phone, taking his eyes off the road long enough for a graceful swerve into the neighboring lane before he finds Enahm's number.

Guttural notes of Dutch flit between them, and it seems that Wim's recommendation is all that's required to get me on the list. "Scott Carney is coming," Wim says emphatically in English. There is a short argument but eventually an accented, "No problem." I smile and realize that I don't really know what I've gotten myself into. That unknown—along with a few thousand dollars—are my cost of entry.

It isn't long until we arrive at the outskirts of The Hague, and the GPS begins to recalculate the route. It recommends an immediate exit from the highway over a bridge somewhere to the north. Hof swerves through traffic, but once we've made it onto the new stretch of asphalt the GPS pushes back into rerouting mode and tells us to make a U-turn. We both scowl inwardly hoping the GPS unit can feel shame. We are supposed to go south now, but the gadget isn't offering an apology. It is precisely the sort of problem we would never have encountered 10 years ago, but even the Iceman finds the convenience of modern technology alluring. "I hate them, but I can't get anywhere without one of these anymore," he says as we spin the vehicle around.

Ten minutes later, Hof parks outside a plain brick building just as the gray sky turns to rain. He opens the door, jumps out, and ducks into a construction site to relieve himself. As he pisses, a thought seems to occur to him, and he calls back to me that before Kilimanjaro I should really meet Geert Buijze, a doctor who had joined him on the expedition up Kilimanjaro 2 years before. Wim says that Buijze knows what I am going to face better than anyone else, and that maybe he could better explain what the dangers of altitude really are. With that, he puts himself back together and is ready to meet the Olympians. It is the same lecture he's given countless times before, but the team, which already bristles with gold medals, listens intently for anything that might give them an edge in Brazil. (And, indeed, it may well have. In the 2016 games, Dutch sailors won two gold medals.)

I sit in the back of the room with my mind wandering to Buijze. As Hof leads the sailors in a breathing session and then an after-lecture ice bath in a blue inflatable birthing pool, I send an e-mail to the doctor hoping for a chance to pick his brain about my upcoming trip to Africa.

The next day Geert Buijze is waiting for me in a swanky hotel on the outskirts of Amsterdam, sipping on a foamy cappuccino. At first I don't recognize him. A few hours earlier I'd taken a tour of his resume online, and the list of accomplishments made me picture a man in the twilight of his career. An orthopedic surgeon with both an MD and PhD to his credit, he's also done stints in the trauma center at Mass General Hospital in Boston and published research on high-altitude climbing, ligament reconstruction, and on the tendency for misquotes in peer-reviewed science papers. In his spare time he climbs mountains. The list, which goes on for a few pages, made me figure that the guy must have at least 30 years on me. So when my eyes lock on an aristocratically handsome man in his mid-thirties, I am surprised to see him wave.

"I just Googled you," he says, pointing to my face on his computer screen. "We certainly have a lot to talk about."

In 2014, Buijze joined Hof and 23 other climbers on a remarkably similar trip to the one I am planning to go on. They summited Kilimanjaro in just a few days without any acclimatization. Initially content simply to be a participant in the crew, Buijze's medical credentials eventually made him a perfect candidate to be the expedition's doctor. Enahm asked him if maybe he would write an academic paper about the experience.

The 2014 Kilimanjaro expedition presented Buijze with an intimidating set of responsibilities. At least half the group suffered from one chronic illness or another. One man had cancer, and there were people with serious heart conditions, chronic arthritis, and an assortment of immune diseases. All this compounded the obvious threat of acute mountain sickness (AMS). Looking for some advice about the actual risks from other mountaineers, Buijze reached out to a mountaineering club in the Netherlands to learn more, but they tried to warn him off. "They told me that someone would die," he says, and then thinks about it. "Actually, they told me everyone would die."

That wasn't quite the sort of encouragement he was looking for, so he looked to the academic literature for a second opinion. An article in the *New England Journal of Medicine* (*NEJM*) analyzed the ascents of 312 hikers, all of whom attempted to get to the top of Kilimanjaro in just 5 days. It was the closest cognate to what his own group would be doing, except that in his case Buijze and his fellow climbers would aim to reach the summit in just 48 hours and to do most of the climb bare chested. Even so, the numbers were grim. Of the hikers analyzed in the *NEJM* study, only 61 percent made it to the top while 77 percent developed

AMS, including a few cases of the most severe form of the condition: hypobaric hypoxia. The mountaineering club's warning didn't seem too far off.

Hypobaric hypoxia is a nasty condition. Although the Earth's atmosphere is 21 percent oxygen regardless of altitude, the lower overall air pressure at higher elevations means that the higher a person goes up, the less oxygen there is available for any given breath. While it can take a long time to die of starvation, oxygen depletion can wreak havoc on the body in mere hours as it shuts down one critical system after another. Typical low-level symptoms include headache, nausea, fatigue, inability to sleep, and dizziness. At more severe stages—at the point where AMS gets a Latin name—people lose their balance and fluid begins to collect in their extremities. Once severe symptoms set in there's no way to reverse the process until the patient descends to a lower altitude.

Absolutely everything in the literature indicated that Hof's expedition was a suicide mission. "I knew that I had a responsibility to warn the group," Buijze recounts. The expedition wasn't just a high-risk affair, it bordered on stupid. The sole thing in the hikers' favor was their faith that rapid breathing along with a half-year-long process of conditioning their bodies to the cold might—theoretically—provide a certain level of protection. If the entire group breathed consciously during the ascent, it might compensate for the low overall oxygen at higher altitude. After all, that was essentially how Diamox worked. The altitude acclimatization drug passively speeds up respiration to counteract the lower saturation of oxygen at altitude. In theory, Hof's breathing method would take the pharmaceutical out of the equation but produce the same basic result. That said, even if the method worked for most people, what would happen if one member of the group fell ill? Would they be able to maintain their concentration and breathing while simultaneously mounting a rescue operation? In a worst-case scenario the entire group could fall like dominoes. As one person attempted to rescue another, they could well succumb to the same environmental factors until the whole team became incapacitated. The prospect gave him nightmares. But on the other hand, what if they made it? If a simple breathing technique could defeat AMS, then there was potential to help climbers around the world who found themselves in trouble at elevation.

So Bujize made a plan. As the first wilderness EMT to ever take on a challenge like this, he designed his own set of safety protocols. His e-mails and messages to the group highlighted their potential dooms, and he started to get a reputation among the other mountaineers for being a worrywart. "I needed to get them to understand exactly how serious this was. It is easy to get caught up in the fun of the expedition and believe that everything would be okay. I needed to scare them a little," he says.

Eventually his imploring prevailed. They all agreed to abide by following a simple buddy system that would make everyone responsible for a climbing partner. He found a checklist of altitude sickness symptoms known as the Lake Louise Criteria that would provide canned ratings and symptoms to watch for during their progress on the ascent.

Since altitude sickness manifests differently in every person, the protocol ranks a variety of likely complications as an expedition ascends higher. Every symptom gets a point value—a mild headache is worth 1 point, severe dizziness ranks as a 3, and every mildly swollen limb on a person's body is worth 1 point (for a maximum of 4 points, one for each limb). Everyone agreed that if anyone scored a 5 they would stop and hold their position on the mountain until the symptoms subsided. If someone scored 7 or more, the two buddies would turn back. After the trip Buijze would use rankings that people recorded and analyze them for his paper.

So while beforehand the crew worried about how they might fare on the mountain, the actual trip was a great success. It has gone down in Wim Hof lore as one of the most profound test cases of the method. A few months after their expedition Bujize published a letter to the editor in the journal *Wilderness & Environmental Medicine* stating that out of the 25 people on the journey, only one experienced severe symptoms. This single man, with a heart condition, had to descend early. Everyone

Lake Louise Score (LLS) for the Diagnosis of Acute Mountain Sickness (AMS): Self-Report Questionnaire

Add together the individual scores for each symptom to get the total score. A total score of 3 to 5 means mild AMS while a score of 6 or more means severe AMS.

Headache	No headache	0	
	Mild headache	1	
	Moderate headache	2	
	Severe headache, incapacitating	3	
Gastrointestinal symptoms	None	0	
	Poor appetite or nausea	1	
	Moderate nausea &/or vomiting	2	
	Severe nausea &/or vomiting	3	
Fatigue &/or weakness	Not tired or weak	0	
	Mild fatigue/weakness	1	
	Moderate fatigue/weakness	2	
	Severe fatigue/weakness	3	
Dizziness/ lightheadedness	Not dizzy	0	
	Mild dizziness	1	
	Moderate dizziness	2	
	Severe dizziness, incapacitating	3	
Difficulty sleeping	Slept as well as usual	0	
	Did not sleep as well as usual	1	
	Woke many times, poor sleep	2	
	Could not sleep at all	3	
		TOTAL SCORE	

else made it to the rim of the volcano. Almost all of them were shirtless the whole time. The 92 percent success rate was unheard of in medical literature. Or, as Bujize wrote in reserved medical-speak in his letter, "The method seems to have a direct biological effect on the autonomic nervous system, and warrants further investigation . . . The remarkable results are most likely explained by the continuous controlled hyperventilation reducing hypoxia severity. However, the resultant state of respiratory alkalosis may cause symptoms such as dizziness and visual disturbances. The fact that none of the trekkers experienced such symptoms is most likely due to the long-term training."

So it worked! It was a huge success, and Hof repeated it a year later with a different group. I am relieved. But Bujize is not willing to assure me that this next expedition will be the same. "We know that it *has* worked, but exactly what factors go into success are hard to say. It is still a very dangerous proposition. It's not like people haven't died following Wim's example before," he adds almost as an afterthought.

One side effect of Hof's sudden international recognition is that tens of thousands of people want to perform their own versions of Hof's feats. Over the years Hof himself has sometimes paid a physical price for his daredevil adventures, but he has always managed to come out the other end alive and smiling. Not so for all the people following his method. In the previous winter two Dutch men started learning Hof's breathing techniques from YouTube videos and found they loved being in cold water. When one particular cold front froze the surfaces of the canals, they challenged each other to swim beneath a stretch of ice. Whereas an emergency diver was on hand to rescue Hof when he lost track of the hole in the ice and his corneas froze while attempting a similar swim, the two friends were neither as well prepared nor as lucky. One of them drowned.

The death cast a certain pall on the Innerfire organization. When I ask him about it later, Enahm says, "We can't be responsible for every extreme action that people say is inspired by Wim. That would be impossible. There are just too many factors." He's right. While it's true that Hof's method helps push the limits of human achievement, nature will win eventually if you constantly test it.

That evening I pilot a tiny economy rental car back to Stroe and wait in the empty house for Hof to get back home. For a 56-year-old man, his schedule seems endless. There are lectures in the morning and evening, halls packed with corporate executives seeking to use his method to train better leaders, as well as nonstop requests from people suffering from debilitating illnesses who see the Iceman and his method as a last-ditch hope. He texted me earlier saying that he was on a special mission out of the country and would be back late.

The night drags on until eventually I again hear the crunch of Hof's tires on the gravel outside. The engine purrs for a few minutes as he sits behind the wheel thinking and breathing consciously to find his center. Eventually he opens the door and I notice that his rumpled suit coat looks even more rumpled than it did in the morning. I want to ask him about how the person died swimming under the ice but it doesn't seem like a good time. He smiles faintly at me.

"I saved a woman's life today," he says softly. After a moment or two he tells me that he was just back from Brussels, an almost 250-mile drive round-trip. There a woman had called him saying that he was her last hope before she took her own life. He figured he didn't have a choice but to at least try to help. "She was hurt—her hips and pelvis were broken as a child and since then she's known only pain. So I spent the day with her and tried opening her up," he says. Hof taught her a technique to dull the pain and retrain her nerves.

"She still has a long journey ahead of her and I don't know for sure," Hof says. "But I think she will be okay."

I am beginning to feel the same way about Kilimanjaro. It is a little more than 2 weeks before we are scheduled to fly to Tanzania, but Hof has other things to think about besides my budding fears about the trip. He shuffles off into the kitchen and looks to scrounge up his first meal of the day. He ponders over some mayonnaise and some pickled shellfish in a jar. The man who spent the morning dancing in a rainstorm in his own living room nods at his meal as if he is saying, "It'll do."

9

PARKINSON'S, BROKEN BONES, ARTHRITIS, AND CROHN'S

IN THE MIDDLE of the night, Hans Spaans growls into his pillow as the cramps slowly overwhelm him. The muscles tighten in his thighs, back, calves, and arms in a relentless charley horse. His body is a hard knot of paralyzing pain, but at least he still has his voice. So he lets an unholy string of words fly from his mouth and into the cotton casing. The cries gain power with every passing second until eventually the sounds of anger and agony fall to a whisper. Somehow, the rage gives him the strength to stumble onto his feet and kick a couple of pillows around the room with the brutal force reminiscent of someone who studied martial arts in their youth. And then, as the anger fades away, just like that, the cramps are gone; he can move again. If Spaans has learned one thing in the 14 years since he was diagnosed with Parkinson's disease it is that sometimes he can swear his symptoms into submission.

It all started back in 2002, when Spaans was still working as an outsourcing consultant for a Dutch subsidiary of the IT company Unisys.

His was a fast and intense life that followed the mantras "I want it all and I want it now" and "The sky's the limit—if at all." At the age of 41 he had no thought of ever slowing down or living a more stable life, and he had no issues about working overtime on projects just so long as they were high profile. He enjoyed the benefits of the bachelor life: driving fast cars, sailing a boat on the lakes and waterways of the Netherlands, and skiing throughout the winter.

That year, on a solo skiing trip in the Banff area of the Canadian Rockies he took a tumble on the slopes. He broke his left pinkie finger and damaged his right hip joint. It hurt, but he figured that it was a minor injury that would correct itself over time. But not long afterward he started to lose sensation in his left arm. Over the next month people said that he looked different, and that his posture was changing. His arm no longer moved when he walked. It wasn't until 2003 that the vague but persistent stiffness in his gait and arm finally drove him to go seek medical advice.

At first doctors thought it might just be a bad case of carpal tunnel syndrome, but they couldn't find anything mechanically wrong with him when they investigated his wrist. A subsequent battery of tests looked farther up his arm for the source of problem, ruling out one explanation after another until finally the doctors determined that the root cause must be somewhere in his brain. Eventually doctors ordered a special scan that required them to inject him with a radioactive dye that would show a characteristic signature of neurological degeneration and decreased dopamine production—one of the only sure signs of Parkinson's disease. It turns out that many cases of Parkinson's begin in a similar way. A small but intense trauma initiates a cascade of stiffness that slowly propagates through the entire body, ultimately leading to progressive and irreversible neurological impairment. Over time the disease can all but stop the interaction between the peripheral body and the central nervous system

That said, there is no one sure path that Parkinson's disease takes for every person. Michael J. Fox and George H. W. Bush both suffer from

Parkinson's but don't seem to have the same set of disabilities. However, as a general rule, Parkinson's progresses from minor motor symptoms and weakness to an all-encompassing disorder that traps a body in a sort of permanent constricting cramp, in which every muscle refuses to relax. The end stages are terrifying; a person can be trapped in their body with nothing but searing pain. Such a fate is possible for anyone with the disease no matter how vibrant they were in their youth, as tragically exemplified in 2016 when legendary boxer Muhammad Ali died after spending his final years mostly paralyzed.

Given his diagnosis, Spaans opted to make the most of the time that he had left instead of dwelling on the future. He would take the drugs his doctors prescribed but wouldn't change a thing about his lifestyle. By throwing caution to the wind he figured that he would just live a little faster and maybe die a little younger than he would have without the condition. And for a while it worked. He popped increasingly large doses of medications that physicians assured him would keep his symptoms at bay, knowing that every dose brought with it an incremental amount of resistance. Eventually he knew that the drugs would lose their effectiveness and that a wheelchair would be just around the corner.

In the early stages, the medication nicely camouflaged the symptoms as it introduced an influx of dopamine into his fraying neurology. But the relief came at a cost. As the number of pills went up, they slowly lost their ability to trick his muscles into action.

Then one morning in March 2011 while he was working at his home computer, his hands just froze in place. Try as he might he couldn't command his muscles to reach the keyboard. He tried to stand up, but his muscles failed him again. Spaans had finally lost control, and his body had transformed into a knot of painful cramps. It was game over.

Later he'd recall the episode almost with humor. "It was funny," he says. "I was stuck on my chair like a statue, like a noodle of pain, and all I could think was that I should call in sick to work for the rest of that day. In truth I should have probably quit everything back in 2004 and focused on myself."

Spaans isn't entirely sure how long he remained stuck, but what came next changed the course of his life. The stiffness and pain washed over him like a wave until it overcame his mind itself. He fell out of his chair. Then, when Hans Spaans the IT worker and playboy was nothing more than the imprint of pain itself, his body shook and shivered with a force that lifted him up off the ground and then threw him back down to the floor. It was as if he was being reanimated by electroshock.

The seizure lasted almost a minute. And then, just like that, the pain went away. Spaans was back in control. He stood up like nothing happened.

The seizure meant two things: first, his working life was over and, second, so was his ability to keep the symptoms of his disease at bay with medication. His doctor wasn't quite ready to give up hope on medical interventions and suggested that he consider an even more drastic hack into his biology—something called deep brain stimulation. In this procedure, a team of surgeons would implant an electrode into his brain that, in turn, would hopefully give him a few more quality years.

It was a tempting offer, but one that he worried would only mask his symptoms further and not combat the underlying problem. As a kid he'd practiced kung fu and yoga. And while he didn't necessarily believe he could cure his condition, he thought that alternative therapies could maybe let him manage his symptoms without as many drugs.

I first met Spaans in Poland, after he'd decided to put his faith in Wim Hof to help heal his crippled body. At the time he'd been only doing the program for a few months and his energy levels were low, but the ice baths and breathing seemed to be having a positive effect. While the rest of us could focus on the training, he could only manage 4 hours every day before retreating back to his room to suffer through the symptoms of his disease. It was intimidating to try to talk to him, as if the effort it took him to form words was syphoning away his energy to do anything else. Still, he saw that there was something in the method that was working.

"Having a condition like this makes you intensely aware of your body," he said at the time. Parkinson's made it hard for his brain to communicate with his limbs. But rather than just endure the suffering, he had started to think of every cramp and every shake as an ongoing conversation with his nervous system. Building on Hof's idea that his body was part of the environment, and that his condition meant his brain wasn't able to send strong signals to his muscles anymore, he tried to work backward and use the environment to send increasingly strong signals from his skin to his brain. If he could just find the right combination of external stimuli that might act as commands, then maybe his brain would begin to listen again. It was a leap of faith, but what did he have to lose? Progress was slow, but he was learning.

"Wim says that the cold is the same as an emotion," said Spaans. And that seemed true to him. Sometimes the intensity of his anger at his helplessness had the power to break through whatever neurological roadblocks were stopping his brain from communicating with his muscles. But anger was also self-destructive. Using it made him hate himself and resent his condition. Cold showers and ice baths generated similarly intense signals, and he didn't need to jump through any emotional hoops to achieve the intensity he needed to make a breakthrough. He could simply plan baths accordingly.

When he started his recovery mission, Spaans tracked his drug consumption and "up-time" on a spreadsheet, and then uploaded it to a Facebook page he called, "Challenging Parkinson's." Within weeks he was able to show that his overall drug consumption was down even as he managed to maintain the same number of good hours each day.

Three years after we first met, I travel to Spaans's home outside Amsterdam to see how he is doing. Living with Parkinson's for 14 years is a feat in itself, so I am not expecting Spaans to have many semblances of a normal life. I predict wheelchairs, ramps, and all the medical accoutrements that come along with managing a progressive disease. When he greets me at the door he confides that he almost had to cancel our appointment because the previous night was a bad one, and he had screamed into his pillow until the early hours. At some point soon he expects the tremors to take over and for the system-wide cramps to set in again. I suggest that we can reschedule, but he waves me in anyway. "Let's just see how this goes," he says.

For Spaans, one small upside of having the disease is that he has found the time to take up a few hobbies. I don't see any of the medical bric-a-brac I'd expected. Instead, his living room is a mixture of Asian art and a collection of guitars. And there is an unexpected masterpiece in the back of his house. When the symptoms set in he challenged himself to build an extension onto his cottage that would be part martial arts dojo and part meditation hall. And this he did. A clay stove serves as a place to burn aromatic herbs, and an Indonesian mask hangs from a stud by a bay window. "I started building it in my free time," Spaans says. "Sometimes I could just hammer in one board a day, but over time it all came together." Even though it is a "bad day," he shows me that he can still kick a sausage-shaped punching bag that dangles above our heads from the ceiling. I am pretty sure that my leg wouldn't reach as high if I tried.

"I can't let this disease stop me," he says before asking if I would like some coffee. I do, and so he brews up a fresh batch as we talk about his morning regimen. Mornings are always the worst; he wakes up stiff as a board and has almost no way to access his limbs. So before the sun comes up, he starts the deep breathing routine that Hof teaches. It usually takes Spaans the better part of half an hour of deep breathing and breath-holds before his muscles start to loosen. When he is limber enough, he takes a cold shower to shock his nerves into consciousness. Think of it this way: Parkinson's disease is a degenerative process in which the brain slowly loses connections with the limbs it is supposed to command, and Spaans uses the Wim Hof Method to strengthen environmental signals to override his failing neurology. It isn't a cure. It's a management routine. In 2015 he is enjoying 11.5 hours of "good time" every day with a lower drug regimen as compared with 2011, when his daily average of "good time" dipped to less than 7 hours. And that progress is more hope than any doctor has given him so far. It's an unorthodox approach—one that his doctor thinks is nothing more than a placebo effect—but at least Spaans can speak about his future again.

Spaans is hardly the only person in Holland using Hof's methods to treat a complex condition. Almost all the people I meet in Holland who are studying the method because of an illness have a similar story. They start out on prescription drug routines that seem to work at first, but then the drugs lose their effectiveness over time. Each turn to Hof as a last resort. While the Wim Hof Method isn't a panacea for every illness, the most startling recoveries I see in the few weeks I am in Holland occur in people who suffer from autoimmune conditions. These are a few of their stories.

THE FIRST THING that anyone notices about Hans Emmink is that he's a large man. He has a big barrel chest, long loping arms, and a gut that leads him comfortably. A 3-inch crocodile tooth hangs from a rope around his neck. Despite appearances, he's also the sort of man who is more likely to embrace you in a bear hug than scowl at you from his great height. In 2012 Emmink didn't stand nearly as tall—in fact, most of the time he was bowled over suffering from the painful ulcers that developed throughout his digestive track. It wasn't just pain. Every time he went to the bathroom, the bowl of the toilet filled with blood. "The first time I saw it I freaked out," he says to me when I meet him near Hof's training center. After the first red porcelain bowl Hans called his doctor, who ordered up a colonoscopy, a procedure that uses a snakelike camera to investigate the colon and intestines. The camera lens captured bright white ulcers every few centimeters. There were hundreds of them. It was Crohn's disease—one of the most common autoimmune disorders in the world today. According to the Centers for Disease Control and Prevention, more than 1.6 million Americans suffer from it. About half of the afflicted will at some point require invasive surgery in which doctors remove whole sections of the patient's severely inflamed and damaged digestive tract. There's no cure for Crohn's, but as with many autoimmune disorders, doctors are often able to keep the worst symptoms at bay with heavy drug regimens—mostly steroids. Limiting or eliminating gluten from the diet also seems to help some people.

For Emmink, Crohn's was a living nightmare. Ulcers started in his mouth and proceeded in a line through his esophagus, stomach, large and small intestines, and into his ruined colon. His joints ached so much that it was almost impossible to work. The doctor prescribed him prednisone, a common steroid, and initially the symptoms abated. He could function on a daily basis, but the relief didn't last more than a few months. He rapidly built up resistance to the drug, and it wasn't long before he was shitting blood again. So, like anyone with access to Google, he searched for experimental and alternative treatments for the disease—there were thousands online—but one then-recent article by Peter Pickkers at Radboud University stood out. Pickkers's study showed that Hof was not only able to suppress his own immune system against a foreign toxin, but the following year the continuing study demonstrated that the skill wasn't only an evolutionary quirk, but teachable. People online were musing that it might be good for any immune problem, so Emmink signed up for the first course he could find.

"It's strange to say it, but I started feeling better from day one," he says. The worst part of Crohn's for Emmink was the near-constant pain, and he found that the breathing alone helped manage those symptoms. As he breathed in more oxygen he imagined that he was breathing out pain, and the mere visualization made it seem to disappear. Immediately convinced that something was happening, he took regular cold showers and meditated every morning for 20 minutes. "I could feel my body fighting back," he says. "It was . . . remarkable." He also stopped taking the steroids.

Fifteen months later Emmink booked another appointment with the doctor who'd found the hundreds of ulcers in the first place. This time there were none. While the doctor was amazed, he wasn't willing to point to a health miracle. Instead, the doctor just gave a shrug and said that while sometimes these things just go into remission, it doesn't mean the disease is gone. The doctor offered Emmink more steroids, just in case the condition returned, but Emmink declined.

"If I had stayed on the meds I would be falling off a cliff all the time. My health would be terrible and I'd be bleeding every day," Emmink says now.

People with stories like this are everywhere around Wim Hof, and sometimes it is hard to make sense of it all. In the span of just a few days I meet at least 10 people who claim that Hof has saved their lives. And yet the skeptic in me is hesitant to point to a specific cause for their miraculous recoveries. As Emmink's doctor correctly pointed out, the medical literature is littered with case studies of people who inexplicably beat the odds. There doesn't always have to be an explanation; sometimes people just get better. This could just boil down to the placebo effect, the remarkable phenomenon where people who are given false treatments can show signs of improvement simply because they believe it might help. There's often no physical basis for the recovery, but recover they do. The placebo effect is so confounding to medicine that truly rigorous scientific studies attempt to control for it if they want to prove that one medical intervention or another is actually better than the mysterious voodoo of the body. Clinical trials set up control groups where one patient gets an active drug and the other gets a sugar pill. If a treatment shows that the drug is more effective than a placebo, then it earns a scientific seal of approval and a place on pharmacy shelves around the world.

And yet while chemicals get mounted on a pharmacological pedestal, it's easy to forget that placebos can be incredibly powerful in and of themselves. As reported by the National Cancer Institute in 2003, in cancer trials placebos can prove to be effective as much as 20 percent of the time whereas leading clinical agents rate at 30 percent. By my math, that means the physical effectiveness of drugs might only be a marginal 10 percentage points over the mystery of self-healing.

And it's not just cancer trials. For example, take the hair-growth pill minoxidil. In the 1990s the world learned about minoxidil under the brand name Rogaine, and it was popularized in the "Hair Club For Men" commercials that played on endless loops on television. In the clinical literature, 63 percent of men regrew hair when treated with the drug, while 35 percent of the placebo group reported similar results. This means that positive thinking, or whatever magic happens when you think you might get better, could account for more than half of the actual hair growing power of minoxidil.

And that's not even the strangest part of the placebo effect. For whatever reason, there's indication that our placebo responses are actually getting stronger. Recent research out of McGill University in Montreal showed that late-stage clinical trials on pain medications were failing because the placebo responses were actually stronger than the drugs. The researchers examined 84 clinical trials for chronic pain published between 1990 and 2013 and discovered that the gap between effective pain treatments and placebos used to be 27 percent, but by 2013 it had declined to just 9 percent. For reasons they still can't explain, there appears to be a cultural component. When the same drugs were tested in Europe instead of North America, the placebo effect was only half as pronounced.

Anecdotal evidence notwithstanding, it's probably not possible to definitively say that the Wim Hof Method is any more effective than the placebo outside of the few studies that have been done in a clinical setting. The people that I meet in Holland might just be success outliers. After all, it makes sense that I'm more likely to meet people who see benefit in the practice than people who abandoned the method after a few days. This of course would skew my perspective toward positive results. That said, some stories are just too astounding not to relay.

KASPER VAN DER MEULEN was running a race when six skittish-looking horses trotted in a pack as they passed him. One of them bucked and kicked its back leg into his body. He doesn't remember the actual strike, nor did he hear the bone in his arm snap. The first thing he can actually remember was the sound of his heart pumping loudly in his ears. When he looked down, he saw a bulge under his forearm, a clear indication that something was seriously wrong. At the very least the arm was broken. Strangely, in the moment, at least, it didn't actually hurt. Enough endorphins rushed through his system that he felt pretty, well, normal. But when the woman who had been riding the horse came to him in hysterics asking if he'd broken something, he knew that he would have to get to the hospital. "I'm not a doctor, but . . ." he started saying while dangling the awkward and crooked limb for her to ogle. She filled in the rest of the sentence herself.

With his good hand he fished his mobile phone from his marathon vest (it was somewhere mixed in with a pair of peanut butter sandwiches he'd packed for the race) and called his wife to pick him up. He sounded calm on the phone. So calm that she knew something was terribly wrong, and she broke into tears before he even mentioned the horse. She jumped in the car and sped toward him. Knowing that she was a few miles away, van der Meulen meanwhile started focusing on his breathing. He wanted to be ahead of the pain when it finally did set in.

Van der Meulen graduated with one of the very first groups of instructors Hof certified to teach his method. Where Hof can sometimes get lost in his own subject matter and often talks himself through endless tangents about winning the war on bacteria or the importance of universal love, van der Meulen has the advantage of seeing the underlying principles of a subject and putting them into plain Dutch (and English). It helps that his first job out of college was as a science teacher. This occupation required him to break down complex ideas into bite-size chunks for middle school students to digest. Before that, however, van der Meulen spent most of his teen years smoking marijuana and playing video games at home. His dietary habits were terrible, and he ate whatever was both available and fast. By the age of 24 he was so grossly out of shape that he could barely run the length of a city block, let alone a mile. His heart beat irregularly and his blood pressure was seemingly always too high. Worse than that, the 240 pounds he packed on his 6-foot-1 frame made him look a little like Humpty Dumpty ready to fall off a wall. Van der Meulen didn't feel well and figured that he was just depressed. When a therapist suggested that he might benefit from a little exercise, he was horrified. It was the first time that anyone had told him he was out of shape. But his therapist persisted, saying, "You know, sound body, sound mind."

So he started making incremental changes to his lifestyle. First with short, breathless runs around the block. Then around the park. He started eating just two large meals a day while fasting the rest of the time. This, he'd read, would help stabilize his insulin production and more closely fit with the patterns that humans evolved with. He also hit the gym. Within 3 years he'd lost 80 pounds. By then he's wasn't just running in the neighborhood but covering 60 miles a week on country roads. Conquering new challenges became a way of life. "I found that as soon as I started doing difficult things that everything else in life got easier," he says. Eventually he found Hof and everything just clicked. The method fit his new persona perfectly, and he geeked out on the biology and budding scientific literature.

He began running marathons and obstacle course races to test himself even more, and it was on one such 20.5-mile trail run that he came across the six-pack of skittish horses. Less than a mile from the finish line and one painful flash of a hoof strike later he was bleeding on the ground and holding his crooked forearm. Eventually the surge of adrenaline began to subside and he could feel fingers of pain creep across his arm and chest. He redoubled his breathing and the rider who was still with him looked terrified.

"You're hyperventilating," she said, no doubt thinking she would need to put a brown paper bag over his mouth to stop him from passing out. Van der Meulen shook his head and asked her to let him concentrate. It took 15 minutes for them to walk to the aid station, and from there his wife brought him to the hospital. By the time he was ready for discharge he was actually kind of cheerful.

An X-ray confirmed what van der Meulen already knew. The ulna—the bone opposite the thumb that connects the hand to the elbow—had fractured cleanly into two pieces. He would need surgery, complete with screws. The doctor told him it would be a long recovery. Furthermore, a large purple horseshoe-shaped bruise stretched over his kidney. A nurse offered him an opium-based pain medication while he waited for surgery, but he turned it down. For him, the broken bone was an opportunity for him to practice pain control. The nurse was shocked. She'd never had anyone in his position refuse morphine, but he did. So, long after adrenaline had blunted his pain, van der Meulen breathed consciously for hours and visualized light moving from his lungs to his arm in lieu of a much easier path to painlessness.

The doctors fitted him with a temporary cast and scheduled a surgery for 4 days later. He spent the evening meditating on his injuries, and when he came in for a follow-up appointment the next day the bruises on his ribs and kidneys were gone. The nurse wondered why van der Meulen wasn't healing like a normal person.

When he finally was admitted to the operating room a few days later, he turned down the drugs again. Doctors gave him only a local anesthetic as they opened up his flesh to set the bone. A few hours later, when he emerged from surgery, a yellowish incision spanned from his wrist halfway to his elbow with broad, inelegant sutures. A nurse told him that he would have to wait for 2 weeks before they could be removed, but asked him to come back the next day so the doctors could have a look at how the surgery took.

Van der Meulen spent the next 4 hours breathing and focusing on his arm. When he tells me this story, I have trouble wrapping my mind around the amount of concentration it must have taken. The effort was constant, but he tells me that it seemed to pay off. When he finally grew tired and went to bed that night, all that was left were the stitches; the resultant swelling from the surgery had markedly subsided. He shows me pictures to prove it. The next day the nurse was shocked.

"Well that went quickly," he remembers her saying. "In about two weeks the sutures should start to itch, and then you can come back and we'll take them out."

Van der Meulen didn't have to wait quite that long. Three days later the sutures began to itch, but when he called the hospital they said it was far too early to remove them. So his wife took them out herself using a pair of kitchen scissors. When he finally came back to the hospital for his scheduled appointment, the nurse examined both of his arms but the injured limb had healed so neatly that she couldn't tell which one had been through surgery without asking van der Meulen. When the doctor looked for himself he shrugged. "You are certainly a medical anomaly," he told his patient.

Yet for van der Meulen, his experience of healing is not so different from those of other people who have studied the method. Nonetheless, his is a go-to story in all of Hof's instructor sessions. At one lecture, as Hof sat in the audience evaluating his disciples, van der Meulen explained how he dropped everything in his life for a few weeks and just breathed and willed his arm to get better. It was exhausting and all encompassing, but the results were real and specific. Hof was almost in tears at the end of the lecture. Not only had he found someone who could help with the growing burden of teaching people the method, he had found someone who understood how the process worked in his own body. It was the same technique that Hof had used when he self-treated his frostbitten feet. "That is *exactly* how it works!" he said to van der Meulen when the talk was over.

The whole ordeal comes with a surprising addendum. Van der Meulen hadn't trained during the time he was focusing on his arm. But about 2 months after it was set he remembered that he'd signed up for a 7-mile race—an event that he would ordinarily expect to finish in the middle of the pack, with a few hundred people both ahead and behind him. He was happy to be running at all, but, wary of pushing too hard and falling only to reinjure his limb, he took his time, focused on his breathing, and took in the view. He expected there to be a large crowd near the finish line, where all the people ahead of him would be celebrating, but when he crossed it he discovered that he had finished in the top 10 percent of the pack. It was his fastest time ever even though he had gone a month without training. He could only assume that the power of his breathing helped push him through.

But how remarkable was his performance, really? As medical professionals will always note, van der Meulen, Emmink, and Spaans are probably outliers. It's the same argument that they made about Hof after he did his experiment with endotoxin. Which is the precise reason why I also had to track down one more of Hof's most notable success stories.

DESCENDED FROM A long line of blacksmiths, Henk van den Bergh is a blacksmith living in an age when most people don't realize that blacksmiths still exist. Founded in 1832, the van den Bergh family forge is one of most notable landmarks in Blaricum, a town that is better known for housing Dutch movie stars than soot and grease-covered men in overalls. Yet van den Bergh has become one of the most notable figures in residence because of his remarkable recovery from crippling rheumatoid arthritis. He's also the founder of an eclectic group of townspeople who spend 20 minutes a day jumping into a nearby lake.

As a child van den Bergh watched his mother become slowly immobilized by the rheumatoid arthritis that ate away at the joints throughout her body. If he was unlucky he might share her genetic predisposition. In the 1980s, when van den Bergh was 24 years old, he took a trip around the world starting in the South Pacific and Australia. He was in the United States, about halfway through the circuit, when his knees began to swell up. Within a few days he could barely walk, and his dreams of continuing around the planet came crashing to a halt because of his ailing limbs. It was the height of the AIDS epidemic, and his friends were sure he'd somehow contracted the disease. Barely able to move and out of money, he flew back to Holland. The pain was so taxing that he spent every day on his back, only leaving his room when his father—a barrel of a man—carried him to a warm bath to soothe his aching joints. His mother had been in much the same condition, her bones eating themselves from the inside until she eventually succumbed to the disease at the age of 57. Van den Bergh's symptoms set in much earlier than his mother's had, and he figured he wouldn't even have that long to live. He contemplated the relative merits of dying early versus living a life immobilized in bed.

"I started to think of my body as a game of Pac-Man," he says to me in his shop. "My bones were those little pellets around the screen and Pac-Man was this disease eating its way through my skeleton." He raises up his hand in front of his face and points to the place on his wrist where most people have a small knob between where the arm stops and the hand begins. There is nothing there. In order to preempt the progress of the disease, doctors sawed away 2 centimeters of bone. "Not that it mattered," van den Bergh says. "It just meant Pac-Man had a new starting point." It wasn't even close to the end of the damage, either. A few months after his surgery the first knuckle on his index finger ballooned to the size of plum. Moving it was impossible and it hurt even when it stayed in place. There was no way he could work in the family business, and as the year marched onward doctors kept recommending the knife; they would cut away the joint and replace it with a dead man's digit.

Rheumatoid arthritis like van den Bergh's starts when the immune system begins to look at joints as foreign invaders. White blood cells and lymph collect in the empty spaces in the joint and disintegrate the bone and cartilage from within. The body literally attacks itself, and the best option a doctor has in such a situation is to offer the patient drugs that completely suppress the immune system in order to save the bone. Unfortunately, shutting down the immune system means living like an invalid with no ability to fight off the ordinary infections of everyday life. As his friends had predicted while he was on his world travels, the condition he had wasn't so far off of AIDS after all.

Van den Bergh speaks to me in a cluttered upstairs office at his shop. We sit at a small coffee table below a low, sloping roof. Beneath a gritty layer of black dust, the walls are covered with news articles about his story as well as pictures of Hof. He points to one in which Hof is wearing a bright orange cap in the middle of an endless expanse of snow. "This man saved my life," he says, "but at first I didn't believe it was possible." He tells me that a friend had heard about Hof's experiments in suppressing his immune system in a Dutch laboratory, and the friend dragged a hopeless van den Bergh to a 2-day workshop in the next town over. He sat skeptically in the back of the room criticizing every word that came out of Hof's mouth. The breathing did not impress him. The cold was infuriating and painful. And he had never been able to do more than a pushup or two in his entire life. When Hof mentioned that the next morning they would all be able to do 30 pushups together without breathing, van den Bergh was done. "Forget it," he said to himself. "I've had a nice day, but I'm heading for the exit now." But when he got up to leave Hof challenged him and asked why he looked so skeptical. Van den Bergh huffed that maybe he should get his money back. But Hof walked up to him and thrust an index finger roughly into van den Bergh's sternum.

"Tomorrow you will do forty pushups," Hof said. "If you don't, you can leave with your money." Hof has a way of convincing people to take him up on challenges, and, being no exception, van den Bergh agreed.

And he never *did* get his money back. By the end of the next day he'd done more pushups than he had ever done before—not quite 40—but close enough to make him reconsider Hof's seemingly impossible claims. A light went off in his brain. He might have even cried a little. If van den Bergh could push his body this far in a single 2-day workshop, what would happen if he did it every day? So van den Bergh dug in. He took ice baths and breathed in the mornings like his life depended on it. Starting at 6:15 a.m. he drove to a freshwater lake outside the village and walked down a small, steel ramp into the water. As he slid in up to his neck in the water he embraced the cold and felt warmth course through his body. "It was like I was connecting with something greater than myself," he says. "Sometimes it was like I could see myself from above as if I was a spirit floating above my body in the water looking down."

After 2 months of regular practice, the pain was gone. A few months later his index finger had shrunk from the size of a plum to a manageable grapelike shape. He could flex it and hold things. His whole body started to feel new again. His morning routine became the talk of the town. A few people laughed when they heard that the 50-ish man was taking nearly naked ice baths in the town lake, but others were curious. One morning someone followed him to the shore. The next day another tagged along. Soon the entire village of Blaricum was fascinated with this blacksmith who was reclaiming his life. The group grew and started messaging each other on their mobile phones, scheduling mass submersions throughout the winter. Before long there were 60 regulars.

Today van den Bergh is a Wim Hof evangelist. And as we talk in the upstairs office he starts getting antsy. His English isn't great, so he suggests that the best way to understand his life today is to accompany him to the lake for a plunge. We walk downstairs past a formidable and ancient forge, then pile into his SUV parked outside the shop. A few minutes later we are close to the lake, but before I can even see the water I hear giggles from three women in bathing suits. Their feet squish audibly into the soggy marsh grass as they jog into view. Van den Bergh waves. He doesn't know them, but he recognizes their spirit. "They're warming up before recycling," he says to me as a matter of fact.

"Recycling?"

"Yes. Going back in. It is much colder the second time."

Sure enough, the ladies pass us again as we make our way out to the dock. They don't slow their pace but simply dive into the water. They gasp as we drop our trousers, don our suits, and join them in the water. The lake is smooth as a mirror. The sun is just beginning to set, which lends the surface a gentle, amber glow. There is no snow in Holland yet, but the water is still absurdly cold. We smile and introduce ourselves to the women.

Van den Bergh takes a deep breath and his eyes gaze out into the distance as if he is a little disconnected, his mind maybe a little fuzzy. "Would you believe that I made it up Kilimanjaro?" he asks. Of course I know that he'd gone up the mountain with Hof. Geert Buijze, the doctor from the first expedition, arranged our introduction. After van den Bergh began learning the method, he said that nothing could stop him from finding his absolute physical limits. He's spent much of his life disabled, but since meeting Hof he's come so far that he ended up shirtless on the top of Africa's highest peak. "Those were the best two days of my life," he says. "No talking, just breathing for two days." He says the group had an almost spiritual connection. When I tell him that I plan to do the same journey with Hof in just a few weeks, he smiles. "You have to be serious about it," he says. And then he gives me a tip. "When the wind rips across the face of the mountain you need to take a mental umbrella and use it to part the air in front of you."

"An umbrella?"

"Just imagine that you're holding an umbrella in front of you like it's your armor for the journey," he says. "I don't know if it actually stops the wind or [if it] just makes you feel like you've stopped it, but it helps. Don't forget to pack it." It sounds like voodoo to me, but then again, what on this journey hasn't?

The water feels good. It burns my skin and dilates my pupils. I want to stay there and enjoy it a little longer. After the first moment of shock, all the endorphins are flowing and the pleasure sets in. The thing about icy baths is that it's only the first moments that are actually uncomfortable. But van den Bergh insists that we get out. The second time will be colder, he promises.

So we rush out of the water, and then we get onto the ground and do just enough pushups to get the blood moving again. I feel my fingers come back to life. He points back to the water and says it is time to get in again. I expect it to feel something similar to the brisk cold from a few minutes earlier, but he is right. Recycling is more intense. Just as my body is beginning to relax from the pushups, the arteries open again and let some blood back in. The water hits me like a bullet and I have a second adrenaline rush. I am high as a kite, but it is also harder for me to relax in the water. A stiff breeze is blowing, and I start to dread how I'll feel when I get out. A shiver wells up from my feet to my forehead. I am out of control, but I stay in the water anyway. It's not like it is going to kill me. After a couple minutes van den Bergh leads me out and we reclaim our garments from the pavement.

"Aaah, Kilimanjaro," he smiles. "You will think about every single breath and it will be the most beautiful time of your life."

Van den Bergh puts his hand up and shows me his once-destroyed finger. The joint bends easily. "The mountain might just restore you," he says.

10

ALL-WEATHER INTERVALS

JUST OPPOSITE A large municipal sewage plant that serves the waste management needs of a sliver of Southern California sits an uninspiring warehouse and office park with just a few cars in its lot. From the outside there is no way to know that this is one of the most controversial sites in American athletic training. A sign on the door warns visitors to beware of the dog, and, indeed, as I approach the glass facade a threatening growl and then a bark halts my progress. I knock on the glass and stand around awkwardly for a few minutes until a heavily tattooed man emerges from a back room, throws open the bolt, and tries to put me at ease.

"She looks real mean, but she's harmless," says Brian MacKenzie of his pit bull as he welcomes me into a pro shop filled with books he's authored, special training masks meant for lung resistance training, and an endless array of other merchandise.

For more than a decade, the 40-year-old trainer has bucked conventional wisdom by proclaiming that the best way to prepare for running long distances isn't to log endless miles on tracks and pavement but instead to focus on skill development, human movement, and shorter, high-intensity workouts. Under his program athletes getting ready for 100-plus-mile ultra-marathons don't run increasingly long distances until they hit their target; instead they train in 1-minute ultra bursts of activity followed by short cool downs. His routines bring a person up to the threshold of their VO$_2$ max—the very point of exhaustion—and can transform a sprinter into an endurance athlete. The method is called high-intensity interval training, or HIIT, and breaks 50 years of running orthodoxy. MacKenzie was an early adopter of HIIT and is one of its chief evangelists.

As a pariah and occasional punching bag for the fitness world, some endurance athletes call him a sham—a guy who is going to make more injuries than world-class athletes. But that hasn't stopped a half-dozen Olympians—two of whom went on to win gold—and countless professional athletes from taking his word as gospel (try as I might he refuses to give me their names, citing their desire for privacy). To be fair, many of the accusations about MacKenzie's programs came about before scientists actually began studying HIIT in earnest. By 2016 researchers at Canada's McMaster University in Hamilton, Ontario, showed that just 1 minute of all-out interval training was *more* effective at getting people into shape than a moderate, 45-minute run.

MacKenzie pats the pit bull on the head and leads me past the front counter and into the back of his converted warehouse. We slip by another room packed with his merch, and pass through a narrow hallway that abuts a corrugated steel barrier and glass porthole into some sort of underwater world. It takes me a moment to realize that I am looking at the interior of a shipping container that has been retrofitted into a swimming pool. The 65,000-gallon container takes up a full half of the facility's available space, while the rest of the gym is crowded with different weights and cardio machines. Sitting on the pavement just outside the loading dock towards the back of the warehouse is an aluminum cattle trough full of slushy ice. He'd gotten it ready specifically for my visit.

Like his friend Laird Hamilton, MacKenzie learned about Wim Hof through the 10-week online course, and he knew he was onto something when he tried breath-hold pushups. The theory behind HIIT is that training at your maximum level of exertion increases your overall athletic ability. In other words, a few 30-second sprints that leave you breathless and near fainting will force your body to access its reserve resources. The short effort is more efficient at improving performance than less intense workouts that remain in an athlete's comfort zone—even if those workouts are significantly longer in duration. When MacKenzie first felt the endurance improvements that come from hyperventilating and breath-holds during pushup repetitions, he realized that breathing could also be used as a tool to help people get even more performance as they approach the threshold of their VO_2 max. And so he began integrating Hof's cold and breathing routines into structured HIIT workouts. On this day he is going to take me through the rounds. He smirks and tries to assure me that it probably won't kill me.

It is a little past 8 in the morning when we sit down and start a familiar breathing routine. We are joined by one of his trainers, a man also named Brian. MacKenzie prefers a regimen of 30 to 45 breaths followed by long holds with loud tribal music filling the studio. We set the timers on our iPhones to keep track of each round. For the last few months I've been recording my times and topping out my breath-holds at about 3 minutes after a few repetitions. But doing it in a group has a sort of amplifying effect. The flitting of colors behind my eyelids seems all the more vivid and my fingers feel cold—a sign of increased circulation. When we stop the communal huffing and puffing, my timer indicates that I've made it to 4 minutes without much effort. The session ends with breath-hold pushups, and that's when MacKenzie signals that the workout can finally begin.

MacKenzie's working hypothesis is that any exercise that makes respiration more efficient will have an immediate effect on cardio performance. The Hof breathing method artificially boosts short-term aerobic endurance, and HIIT routines can take advantage of that sweet spot.

He leads me to something he calls an assault bike, a sort of stationary bicycle with a fan for a front wheel and two moving handlebars. It doesn't look too intimidating. Nor, at first, does his command to warm up for 3 minutes. A digital display on the bike shows me that a comfortable pace for me generates about 300 watts of energy.

"Now give it everything you can for a minute," he says, urging me into a sprint. So I push against the pedals and handlebars like a horse who has just felt spurs for the first time, trying to call up every ounce of power my muscles can offer. For 15 or maybe 20 seconds I feel superhuman as the wattage jumps to a point just shy of 1000 watts. But then, at about 30 seconds, I've burned through my initial stores of energy and start to fall off. Hard. The wattage dips low, and there is nothing I can do to stop its descent. I want to stop altogether, and finishing the bottom half of the minute is pure torture. All told, my average sprinting wattage was a relatively pathetic 500. Suddenly the assault bike doesn't look so innocuous. My heart is racing in my chest, my throat starts to close up as if it is contemplating taking up asthma for a hobby, and when I dismount I think I might fall over. At least I've found my wall—in just 1 minute I've exhausted my entire wellspring of endurance. Hell, I sort of want to go home. But MacKenzie assures me that I am nowhere near done.

As I get my bearings back MacKenzie leads me to the ice trough and tears off his shirt, revealing a body rippling with muscles. A tattoo of an owl spreads across his entire back. He says it's an homage to the artistic traditions of the Native Americans of the Pacific Northwest whose connection to nature and relentless courage and endurance have been guiding forces in his quest to create the perfect athlete. In that moment the words of Rob Pickels from the Boulder Center for Sports Medicine echo in my mind: I will never be a world-class athlete. I wonder why I am even trying to keep up. Then he plunges into the ice, going from the 90-degree California heat into its diametric opposite. He lets out a gasp, then he focuses until he is calm beneath the water. After 5 or 6 minutes he gets out and lets me take my own turn.

I am still flushed red and no doubt running the equivalent of a high fever after my minute on the bike. I figure the ice will bring my body heat down to a normal level almost instantly. But when I get in, the temperature shock triggers vasoconstriction in a way that I have never experienced before. Hot blood is already coursing through the arteries in my extremities as a result of my workout, and when they clench shut with the autonomic programming designed to preserve core heat, a surge of heat pushes back into my core. For a moment my core temperature rockets even higher. I grow woozy, and I whine to the photographer, Chris DeLorenzo, who has come along that I don't feel very well. My legs cramp up and I try hard to relax and straighten them along the bottom of the aluminum trough. It takes half a minute to get under control—the dizziness goes away and all of a sudden the intense pain of the ice transforms into soothing relief. Endorphins buzz around my brain, and I relax my neck against the wall of the tub. It is a moment of pure bliss. After 5 minutes I put my soaking feet onto the hot pavement. I feel almost perfect. The ice countered the debilitating effects of the HIIT. I am ready to get back on the assault bike.

"The beauty of the ice is that it forces you back to your center," says MacKenzie, as he watches my transformation. The next stage of the workout involves getting back on the bike, but this time I'll do 30 fast, deep breaths before hitting the pedals. Blowing off CO_2 and taking in oxygen makes the pushing feel easier. This time, it takes 45 seconds to hit my wall. Not a bad boost for under an hour of training. We do four 1-minute assault bike intervals followed by hyperventilation, and then we do one without the breathing technique. There is an obvious difference when I don't prepare my breathing. It is like I am back where I'd begun, near the verge of passing out. In the last interval I am not supposed to prepare at all. Instead MacKenzie tells me to do the power breathing *while* pushing the bike as hard as I can. Later I'll discover that I've forgotten to record the wattage, but I remember it as being the easiest rep yet.

After flitting between sprints and hyperventilation during an hour on the bike I am ready for a rest, but MacKenzie still plans to show me his pool routine. Kept at a steady 70 degrees, the water is refreshing for someone just coming off a workout. But it is cold enough that when Chris, the photographer, dives in with us he begins to shiver in just a few minutes. He has to put on a neoprene top in order to hold the camera steady. The small swim tank made out of a shipping container is a touch smaller than the cabin of a school bus and is just the right length for a few short laps.

As it turns out, this tank is where MacKenzie helped develop XPT in conjunction with Laird Hamilton. Although they live on opposite sides of Los Angeles they both found the combination of using dumbbells in the water—something Hamilton pioneered eons ago—and a modified Hof breathing routine to be a good way to push their physical limits. MacKenzie leads me on a few familiar laps in the pool while holding different amounts of weight. On one underwater lap I carry a large, black medicine ball along the bottom. In another I swim across the surface burdened by a 50-pound weight, and I'm barely able to keep my head above water. Pool training like this helps condition the body to be more efficient with oxygen. Indeed, just being in the water triggers all sorts of physical reactions, first by lowering the heart rate and second by increasing the atmospheric pressure on the skin. It also lowers the impact of every exercise, putting less strain on joints and muscles while still maintaining a high workload.

After another half hour in the pool, MacKenzie slips below the surface of the water and bleeds the air from his lungs in a steady stream of bubbles. He sinks to the bottom of the container and lands cross-legged on the floor. He opens his eyes and stares across the short expanse, once again finding his center. When he's located it he rises gently up to the surface and takes a deep breath.

MacKenzie is still tweaking his routines, but it is clear that he is onto something. He keeps track of the various exercises on a dry erase board on the wall opposite the pool. "I'm doing this because it works, not because I have a grand idea about an underlying theory of evolution," he says. One day he sees a future of gyms across the country with cold chambers, assault bikes, ice baths, and swimming pools all working together to help add environmental conditioning to any serious athlete's workout routine. I am exhausted, but the idea feels right. It might not be long until there are XPT centers facing off against Bikram yoga studios and CrossFit centers across the United States and Europe. And, yet, part of me also wonders if most people really need perfect optimization to achieve a good workout. Indeed, some of what XPT offers already exists in a stripped-down version in the form of a tribe of athletes that started on the East Coast who are committed to working out no matter the weather.

IN THE WINTER of 2015 a series of blizzards blanketed the northeast United States. The storms left Boston so far under the snow that most people only had just enough time to dig themselves out before the next blizzard came and buried them again. The snow made it impossible for commuters to move around the city, and the snowy banks outside the tenements and brownstones rose so high that some people made a habit of jumping out of their third story windows into the cushiony fluff. Enough people ended up in the emergency room as a result of those stunts that the mayor had to publicly chide the city's foolhardy inhabitants for their apparent collective death wish. As the city grappled with the emergency, one small group of dedicated athletes who'd made promises to work out without any concession to the weather had a different plan. They showed up at Harvard Stadium on the night of the storm with snow shovels and salt to dig out as many of the 1,147 concrete stairs as they could. They had to prepare for the following morning, when more than a hundred athletes planned to run up and down the sections until they'd completed a lap of the horseshoe-shaped coliseum, a feat they refer to as the Full Tour.

November Project is a free fitness movement that is unique in a world of elite gym memberships insofar as all that it takes to join is a pair of sneakers and a diehard desire to get outside and start sweating. It was founded by two veterans of Northeastern University's rowing team who missed the days when their workouts on the Charles River let them eat 7,000 calories a day without gaining a pound. But after college they noticed that every gym and boot camp in the city charged a premium for a chance to sweat. So in 2011 the friends, Bojan Mandaric and Brogan Graham, made a pact to meet every Wednesday before work to either run the stairs of the stadium or make up their own routine of sprints, burpees, and pushups in a nearby park. They would track each other's progress, set goals for the number of sections they completed, and encourage themselves to push just a little harder than they did the week before. They created an online calendar with the title "November Project" for no other reason than their first workout happened in that month. One of them spray-painted the words onto a T-shirt and when someone asked what it was, he told them to show up to a workout. Soon there was a Facebook page, a Twitter feed, and a mention from the official Twitter account of a local athletic outfitter called Marathon Sports. In a matter of months word spread across the university town that there was something special going on in the mornings at Harvard Stadium, and the outlines of a movement started to take shape. At first, curious participants came by twos and threes, but before Mandaric and Graham knew what was happening good days brought out 300 or more people. Now, November Project is spreading across the country like a virus and, last time I checked, was in 29 cities. By the time you read this that number will likely be much, much higher.

November Project is now responsible for a continent-wide code of weather invulnerability. Is it minus 30 in Toronto? Whatever, it's time for jumping jacks. Is there a hot wind blowing in from Mexico into Los Angeles? How about some pushups with claps in between? Freezing sleet in Milwaukee? Please. Each city identifies as its own distinct tribe, but all share the same basic spirit of joy and commitment. I'd worked out with the tribe in Denver a few times, but I wanted to come to Boston to see where it all began.

My breath puffs out a white cloud of mist as I make my way to lower-Allston in the waning days of October. I park my rented Toyota

Corolla outside the stadium and emerge at the bottom of the horseshoe. It is half past six in the morning, and I am running late enough that there are already several hundred, maybe even a thousand, people in the stands by the time I make it to the gate. I drop my bag at the starting line on the far north side of the stadium, where a small group of volunteers are using black spray paint and a stencil to mark the November Project logo on whatever random athletic gear people drop off. This is how the tribe brands its now ubiquitous Grassroots Gear. The crew works with machinelike efficiency—dropping the stencil, ejecting a cloud of black paint into the air, and moving on to the next one. There are at least a hundred shirts to get through before the hour is over. I doff mine, tighten up my laces, and head out onto the concrete with the hope that I can at least catch up to the back of the pack.

In front of me is what looks like a swarm of ants wearing neon sneakers and bright, form-fitting shirts. Michael Jackson's *Thriller* booms around the stadium from an iridescent DJ booth. It is an appropriate selection for the week before Halloween, and I marvel as a line of women take a break from their stair sprints to lurch like zombies with moves ripped straight from the video.

The day's workout is special; it is what they call a PR, or "personal record," day. Every month runners pit themselves against sections of concrete seats that are roughly three times the size of ordinary stairs. Then they jog back down to the bottom, move over one more section, and repeat until either 35 minutes have elapsed or they've completed all 35 sections. It's a tall order for an average athlete, but when I ask a man carrying a baby on his back where the founder is, he points to a small, neon green ant on the other side of the horseshoe. "That's Bojan," he says. He's covered the whole distance in what couldn't have been more than 20 minutes. What's more, I am pretty sure that it is his second time through the course, as he'd led the 5:30 meet-up as well. I sigh, resume my climb on section seven, and have no doubt that my thighs will be sore tomorrow.

The workouts vary—something that I'd already experienced with the

Denver tribe on the steps of the Colorado State House. On ordinary days volunteer organizers select a pattern of burpees, sprints, pushups, jumping jacks, and workout-themed card games for whoever shows up. There are also tribe-specific exercises like "hoistees," "Sebastians," and "Bojans" that are specific to November Project. Every week has its own theme. But whatever the theme or exercise, the real driving force is that you are working out with hundreds of other people, cheering each other on when someone starts to flag, and dispensing an occasional hug. No one is judging anyone else's performance. It's a nice mix of community and athleticism.

Across the field an impossibly fit man in a neon green warm-up jacket throws his arms wide and does a dead sprint up a section of seats while simultaneously shouting, "I'm a bear! I'm a bear!" Without any hesitation another equally impossibly fit man in a neon orange jumper and headband crows back, "That's not what a bear sounds like," and proceeds to growl with his best grizzly impersonation. Then they give each other high fives and fall to the pavement performing synchronous one-legged burpees. It is a bizarre sight, but one that fits into the absurd atmosphere of the moment. There's a good chance that they don't even know each other.

All this is great, but it's not the stair work or eclectic programs that brought me here. What fascinates me is the mental determination that everyone in November Project agrees to maintain. They're as dependable as the US Postal Service: rain, shine, or snow they'll be outside pressing their own personal limits. Every time of year offers its own weather challenges and, while I doubt most people in November Project think of it this way, working out in the outdoors like this every week puts their entire tribal biology in touch with the seasons. They're not only getting in an hour of cardio, they're also telling their nervous systems what time of year it is and activating unconscious evolutionary processes ingrained in their very genetic blueprints.

Weather triggers specific neural circuitry to get ready for whatever season is approaching, building brown fat for the winter and opening sweat glands and increasing peripheral blood flow in the summer. Though it is only October, the temperature is just barely above 40 degrees when the workout started that morning. Bare-chested men and ladies in sports bras sweat up and down the stairs despite the chill.

In about 25 minutes I've made some good headway on the circuit of stairs and start to pull into the final stretch. As I kick up the last of 35 sections of concrete steps my legs feel like jelly beneath me, yet somehow I hobble to the finish. A small group cheers me on and an athletic blonde lady who I'll never see again gives me a hug to congratulate me for the effort.

At the end of the course I finally meet Mandaric in person (his partner, Graham, has recently relocated to San Diego). Like the two men with the bear impersonations, he's a tall, impressively fit guy. He sports a shaven head and a big, bushy beard. Sweat streams off his forehead, but after sitting by watching the final few hundred people cross the finish line he is starting to feel chilly and begins rubbing his hands over his bulging biceps. I ask him if he is surprised that a simple fitness pact he made with his friend has grown so large. He smiles and says that whatever it is that November Project offers, it's something that people must be missing in their daily lives.

"In the last couple hundred years we've put all these barriers on ourselves." A native of Serbia who nevertheless speaks immaculate English, he starts to impersonate the vast majority of the city of Boston by throwing on a Southie accent. "They're like, 'Oh fuck! It's cold outside. I'm just going to sit at home and be comfortable.' They don't realize that the entire human race has been conditioned to think that the outdoors is dangerous. Or that working out in the cold is lunacy. But guess what? This is something that people have been doing for hundreds of thousands of years. We were made for it."

In the last few months November Project has grown successful enough that Mandaric was able to quit his day job and dedicate himself full-time to the various tribes (a generous sponsorship deal with The North Face also helped grease the skids). I ask him about the cold, and if there's ever be a moment when the weather would be harsh enough to merit the cancellation of a workout. Mandaric wrinkles into a full-facial "no."

Then he starts to tell me of one tribe up in Edmonton, Canada, where it gets down to negative 40—the temperature where Fahrenheit and Celsius meet. "We have thirty-five people who come out every week, and the colder it is the bigger a badge of honor it is to show up for the workout. In fact, up in Edmonton anyone who comes out on a day where it is minus thirty or below, we give them a literal badge. Once a kid came up there and missed it by a single degree—it was minus 29—and we turned him away. It just motivated him to come back the next week," he says.

Pushing the limits of endurance and overcoming the weather is part of November Project's allure. The kid in Edmonton wanted to accomplish something significant, something that sounds dangerous. And, more importantly, something that any human *can* do if they set their mind to it. "We do this now because it's fun, but we used to do it because we had to survive," Mandaric says.

For most of human history the option of coming out of the cold into a comfortable home was never on the table. At the very root of adaptation is the struggle for survival. MacKenzie and November Project both grapple with how far they can safely push human endurance in extreme environments. They use that knowledge to train athletes and help them find their inner human strength. But every person has their own limits, and if they cross that line they give nature an opportunity to collect a fatal toll. Sensible athletes like MacKenzie and Mandaric pull back far before then. Only people who routinely flirt with this fatal razor's edge ever truly know what the human body is capable of. And to understand those limits, I have to look past athletic training and into the protocols that prepare someone to actually face death. I need to understand how soldiers train. And that investigation takes me back almost two centuries, to when the weather both won and lost wars.

11

COLD WAR AND THE VITAL PRINCIPLE

PIERRE JEAN MORICHEAU-BEAUPRÉ sat proudly on horseback somewhere in the middle of a column of men that stretched across the horizon. It was the summer of 1812 and the regimental surgeon was part of one of the most successful military forces ever assembled. Napoleon's Grande Armée numbered close to 700,000 men as they departed France, ostensibly on a mission to free Poland from the reign of the Russian tsar. As is often the case with invasions of Poland, the actual goal of liberation was, in reality, a ruse to ease political dissent. Napoleon's true aim was to crush Russia and redirect control of the empire's resources back to Paris.

The newspapers of the day predicted a swift conquest for Napoleon's charges, which were fresh off of a string of victories. And indeed it started off that way with the beleaguered Russian army fleeing east at every opportunity, allowing the French force to advance deep into the heart of the country. They reached Smolensk by September and continued on to Moscow almost unmolested. Just outside the capital city the tsar staged a fitful defense in a battle that killed almost 75,000 soldiers. But the victory wasn't decisive, and as the French marched to Moscow acrid clouds of smoke billowed into the sky. Beneath the haze, the city glowed like an ember.

Aware that they would never be able to defeat the French in a direct military engagement, the Russians opted for a different strategy and burned every field, farmhouse, and granary they had once vowed to protect. The great wall of flame torched whole city blocks, and the inferno consumed the once-majestic metropolis. When the fires finally abated all that was left were a few smoldering ruins.

Until now, the Grande Armée's success had come from its ability to travel fast and light. It quickly overwhelmed enemy defenses as it cannibalized its rival's supply chains like a parasite eating its way through a host. Napoleon observed that every "army marches on its stomach," and without a cumbersome supply chain behind them his soldiers survived by looting the countryside they conquered. This meant that victory was their only option. Stopping even for a few weeks could mean disaster.

What Napoleon hadn't predicted was that there might not be anything to capture once they arrived in the Russian capital. The scarcity of supplies forced an almost impossible decision. They could attempt to overwinter in Moscow with no provisions, or instead make a sad and embarrassing march back to a base in Germany where they could restock their supplies and prepare for another campaign. Unable to decide the best course of action, the army lingered in Moscow for the entire month of October. Then in November the temperature plummeted to positively Siberian levels. By then the men of the Grande Armée were starving, and the only option that remained was to order one of the most miserable and calamitous retreats of all time.

As a surgeon, Beaupré's job was to help oversee the health of the army with the crude tools of the age. In times of combat that meant amputating limbs and providing palliative care for ailing troops. In retreat it meant simply witnessing the horror of a once-invincible army disintegrate before him. He didn't have much time to pause on the march to help anyone in particular. The best he could do was record his observations of how soldiers perished in the frigid snow.

"How many soldiers have I seen wounded or sick, retreating without order, at random, sad, pale, and dejected, begging with tears in their eyes a morsel of bread from everyone they met on the road?" he asked rhetorically in his tome *A Treatise on the Effects and Properties of the Cold: With a Sketch, Historical and Medical, of the Russian Campaign*. From a clinical perspective the military folly was an opportunity to understand in depth how the cold can destroy a person: either quickly overnight or over a long period of stress and deprivation.

Safety was 750 miles away when the column of soldiers turned back toward home. Meanwhile, mounted Cossacks regrouped and traveled along a parallel course, harrying soldiers who left the safety of the road to scrounge for food and supplies from the countryside. Isolated by their own hubris, soldiers froze by the score. In their slow deaths, Beaupré came to believe that physical limits were not necessarily the most critical factor in determining who lived and who died. He believed that mortality hinged on mind-set. He noted that the Italians, Portuguese, and Spaniards in the army grew melancholy at their condition. Hailing from more temperate parts of Europe, those soldiers "obliged to brave an austere climate unknown to them directed their thoughts towards their country. Unfortunate however was he who weakened himself further by abandoning himself to somber and too discouraging ideas and reflections! He was sooner seized by the cold, and he prepared or hastened his death," Beaupré wrote. Surely, soldiers raised in warm climates would be less immune to relentless cold, but importantly Beaupré saw that the soldiers who lost hope were the first to fall. He resorted to calling this ineffable quality that animates life the "vital principle," which he linked to the body's ability to generate warmth and resist the environment. "Life is a lover of heat. He that is asphyxiated by cold can look for salvation in heat alone," he wrote.

By the end of November the retreat was in shambles. A steady, cold north wind left soldiers' faces vulnerable, and Beaupré witnessed men's eyeballs freeze into blindness. Entire regiments disbanded as men froze in their footsteps. He rode past countless troops who, in their last moments of life, threw open their coats and stripped themselves bare. In their final confused moments, as hypothermia closed in, the nerves in their body could no longer identify the sensation of cold and instead signaled overwhelming heat.

Those lucky enough to collect wood for a fire and catch a few hours of sleep discovered that even the warmth of flame hid danger. When the exhausted men basked in the momentary comfort they fell into deep slumbers, only to leave their fires untended. When the flames went out their relaxed bodies couldn't muster a new round of inner vitality to start up again. The first warm fireside hours of sleep were delicious, but "Far from finding safety in the sweets of sleep, they were seized and benumbed by cold, and never saw daylight more."

Somehow Beaupré survived the first leg of the trip to Smolensk. The mercury in the surgeon's thermometer wavered horribly between a low of minus 4 degrees Fahrenheit and a peak of minus 16 degrees. By then the army was little more than a collection of roadside graves piled 20 bodies high. Those not dead yet groaned in agony and delirium, making easy targets for lance thrusts from marauding Cossacks wishing to ensure that no Frenchman ever returned to their country. Beaupré offered what remedies he could to men who had fallen to frostbite. Many of the treatments were little more than home cures informed by a misguided belief in Greek humors and folk traditions. As fingers and toes froze and blackened, Beaupré applied packed snow on the injuries and rubbed vigorously. For those gripped by frostbite the ice generated a phantom sensation of warmth similar to the misguided sensations that made some soldiers strip in their final moments. Unfortunately for his patients, ice rubs do nothing to reverse or provide relief from frostbite.

By the time the Beaupré reached the city of Vilnius in present-day Lithuania—one of the first cities that Napoleon conquered the previous summer—there was no fight left. Only 30,000 soldiers remained out of the almost three-quarters of a million who had originally set out.

It wasn't the first or last time that the weather bested one of the world's most formidable fighting forces. Throughout history far more casualties have fallen to the insults of weather than they have to enemy action. A brief tour of military tragedies includes the events of 218 BC, when a Carthaginian commander named Hannibal Barca tried to bring 38 African elephants over the Alps with the intention of foisting a new weapon of war on an unprepared Rome. The march cost him half of his 46,000 troops and all but a handful of his pachyderms. In 1242, Teutonic knights had a go at invading Russia while wearing full-plate armor. Under the command of King Guy of Jerusalem the knights tried to sack Novgorod. Unfortunately for Guy, when the armies clashed on the surface of a frozen lake, almost 400 metal-clad fighters froze to death when they fell through the ice, effectively ending the eastward expansion of the crusader kings. In 1742, when a different French army retreated from Prague across snow-covered mountains and ravines, they lost 4,000 men in 10 brutal days.

A little more than a century after Napoleon's doomed campaign, Adolph Hitler took his *blitzkrieg* across Poland and marched 3.2 million soldiers toward Moscow without outfitting them with proper winter gear for their invasion. Placing too much hope on the probability of a quick victory, the march ground to a halt in the snow. Halftracks and tanks froze, weapons malfunctioned, cases of frostbite soared, and soldiers froze to death in their trenches. Perhaps fortunately for the rest of the world, nearly a million Nazi soldiers died that winter.

The lesson provided by these examples is simple. Wars are fought as much with the weather as they are between men. And for every story of defeat in winter months, there is often also an equally potent story of triumph by victors who were able to deal with the cold even marginally better than their enemies.

Potential conquerors take note: The routine and extreme stresses that soldiers encounter in the field might be more important than their ability to kill. But how does an army truly prepare itself to push the limits of human endurance when the elements are against them? It's a problem that the US military has had to relearn with every theater it engages in. Proper supplies are, of course, vital, but most modern training programs aim to build on Beaupré's "vital principle." But as with many lessons, the road to progress for the American armed forces began with a tragedy.

It was winter in the Florida panhandle. Second Lieutenant Spencer Dodge was wading through the soupy, spring-fed brine of the Yellow River. An icy fog rolled in over the swamp and the 60-pound pack he carried on his back pushed him down until the water came up to the middle of his chest. Eight sergeants barked orders from the riverbank for Dodge and his fellow soldiers to string a guideline across the water to a hummock of sweet gum and cypress trees about half a mile away. The job would take 6 hours. It was February 15, 1996, and Dodge was in the final week of one of the Army's most hallowed training programs—one that, if he completed it, would enter him into the storied ranks of the Army Rangers. The course aimed to identify a soldier's breaking points so that only the toughest would make it through. Over the several weeks of training they sweated in deserts, hiked over mountains, and now they were soaking in an icy river. Every day was a new challenge and every man in the unit was the definition of fatigue: hungry, emotionally drained, and physically exhausted. Every one of them started the training with 40 more pounds of fat and muscle on their frame than when they ended it.

Dodge had seen 161 men from his class of 334 wash out. A few days earlier one would-be Ranger went incoherent from being in cold water too long—one of the first stages of hypothermia. When a sergeant ordered the unlucky trooper to touch his nose, the confused man instead touched his chest. The mental slip meant he didn't make the cut. Dodge wasn't about to make the same mistake. So when the water sloshed over his body he refused to complain even though he couldn't control his shivering. Perhaps his mind had been so hardened to deprivation that he had no bearing to understand that his body was reaching its limits. In time, though, the telltale signs of hypothermia set in. He lost the ability to focus on the task in front of him. He grew irritable. And his body temperature plummeted so fast that even the convulsions of his muscles couldn't make him any warmer. He was still chest-deep in the water when his heart stopped.

The training exercise continued amid a din of walkie-talkie chatter and signal flares as a few soldiers tried to resuscitate the fallen man. A helicopter from the nearby base hovered over the river as it tried to retrieve Dodge's corpse. As the blades sliced through the air above them it inadvertently pelted the still-waterborne Rangers with an icy spray. Many who were already shivering fell into steep declines. By the time the leadership realized the magnitude of the problem, four would-be Rangers were dead, and many more were in critical condition.

It was the worst night in Ranger training history since 1977 when, in an eerily similar incident, two other candidates died in the same spot in the Yellow River, also of hypothermia. The Army knew something about the potential dangers of the cold, but the pattern of deaths among elite troops was enough of a shock that the top brass wanted an explanation of how the program could go so wrong. And yet, beyond the tragedy of the lost soldiers, the fatalities invoked an interesting dilemma. On one hand the military wants to tap into whatever reserves of human fortitude exist within a person, but the only way to discover how deep those reserves are is to try to exhaust that soldier and bring him perilously close to their own mortality. Army Rangers don't get to choose the environment they will fight in, and there's good reason that the military would want to know what the gap is between tolerable suffering and certain death. They needed more than a hazy idea of a vital principle. They wanted the crisp answers earned from scientific investigation. To conduct the necessary research they tasked a young Army captain with a PhD in human physiology to review and update all the recent literature on soldiers in the cold in order to prevent similar incidents from happening in the future. The captain's name was John Castellani.

At his disposal was an almost 50-year-old facility at a military base just outside of Boston in Natick, Massachusetts, called the United States Army Research Institute of Environmental Medicine, or USARIEM for those inclined to employ unwieldy military acronyms. The institute is part of a larger military campus called the US Army Natick Soldier Systems Center, which is most famous for developing the oft-complained-about field ration called the MRE, or Meals Ready to Eat, that every soldier carries into battle along with ammunition and body armor. In fact, if a soldier depends on it in the field, it's pretty likely that it was tested first in Natick. For a military whose stated mission is to be able to wage two simultaneous wars in two different theaters anywhere on the globe, there is no question that the $18 million annual price tag for the facility is a bargain if it means keeping soldiers and their equipment performing at a high level in every environment. After all, it's not just the soldier firing a gun on the front line who might succumb to the elements. The withering effects of weather have the potential to impact every aspect of training, logistics, transport, and even just hurrying up and waiting for orders at base.

The most important facility on the campus is a series of chambers that are designed to mimic every environment on Earth. There are rooms that can produce a constant deluge of rain or that can ramp up the heat to almost unbearable temperatures with searing 250-watt heat lamps. At the flip of a switch the same basketball court-size room can start a 70-degree drop with the aid of refrigeration coils and a wind turbine. There is also a water immersion facility and an altitude chamber that can mimic the barometric pressures atop Mount Everest. Most of the time the test chambers are used on inanimate objects. A camo-clad researcher might order an artillery piece to hang out in a tropical rainstorm for a week to see which parts rust first. Or maybe they place a few crates of MREs in a heat chamber to see if the plastic packaging degrades. There's even a room on the base equipped with a flamethrower, presumably to simulate how everything from Humvees to parachutes survive actual fire. Whereas most of the base focuses on everything a soldier might need outside of their skin, only Castellani and a handful of other scientists at USARIEM investigate what happens to soldiers from the skin-in.

I visit the research center in October 2015, around the same time I shake hands with the cofounder of November Project, in order to meet with Castellani and observe his research. The lab is the intellectual descendent of a 1927 program at nearby Harvard University called the Fatigue Lab, which was started with the mandate to understand how heat, cold, and exhaustion affected workers who spent most of their time in extreme environments. The Harvard research helped develop protocols for exposure that the military used throughout World War II. The Harvard lab generated detailed tables that calculated heat loss by length of time and type of exposure, and in 1995 the Army was still relying on the same body of research to design their protocols. But the spate of deaths made Castellani wonder if perhaps they had missed something.

"Those earlier experiments were done on soldiers who were relatively happy and healthy," explains Castellani. "They didn't account for the extremes of Special Forces training, the lost muscle and fat mass and fatigue that are the new standards." His first order of business was to redo the original tests with a different group of freshly minted Rangers who had just completed the program.

When he started to study hypothermia, Castellani put his cadets right back into simulated environments that mirrored the conditions where their colleagues died. Over the course of several years, scores of cadets fitted with rectal thermometers marched on treadmills in his cold chambers as he painstakingly measured their heat loss until their core temperatures were 95 degrees. He dunked soldiers waist- and neck-deep in 50-degree water and had them walk on a special, submerged treadmill to simulate the work a Ranger might do during a river crossing. As they walked in place, Castellani recorded the steep thermal decline in their bodies as the heat they generated from exertion dissipated in the cold water. All but two of his volunteers washed out of the study by 4 hours, and none came close to the 6 hours the Rangers in Florida were ordered to toil in the water.

Following the mandate of the top brass, Castellani refined and amended the cold-water tables to align better to the needs of soldiers in training, and then went on to calculate other ways that the climate changes the battlefield. He helped elite units figure out how pressure and speed of insertion affect soldiers on a mountain top, and if diet could predict how many people in a group might come down with the dizzying and possibly lethal condition of acute mountain sickness—the same condition that I will potentially be battling on Kilimanjaro.

His research has powered apps for military cellphones that commanders can use to predict how their troops might react under harsh conditions. "It's the sort of thing that a leader could use to know in advance how many of his troops might get sick," he says. Altitude, it turns out, has a deleterious effect on unit cohesion. If inserted via helicopter or parachute from sea level to 14,000 feet without acclimatization, a soldier might grow withdrawn, aggressive, and uncooperative, qualities that prove disastrous if trying to coordinate high-precision missions. Of course, the tables aren't prescient enough to know how a specific individual will perform under environmental duress, but they can tell a lot about how a group of people will fare overall. And, in some ways, aggregates are more important. Castellani explains it this way: If a commander "knows that 25 percent of his command is likely to be incapacitated from altitude, then he can bring 25 percent more people to accomplish the mission." In a recent field test in which an Army unit rapidly summited Alaska's Mount Denali (at 20,308 feet it is the highest peak in North America), the app was spot-on in predicting about how many people would get sick and have to be brought to a lower altitude to recover. "It made my boss very happy," Castellani says.

Certainly the very limits of human endurance are always on the radar at USARIEM, but to be successful, armies also depend on the multitude of smaller, more mundane things that happen far from the front. When I arrive at the institute, I learn that Castellani is ramping up a new project to test soldiers' manual dexterity in cold temperatures. Long before hypothermia or frostbite sets in, a person experiences a slowdown in fine motor skills. This slowdown is one of the first signs that a person is getting cold. Loss of manual dexterity might mean that a soldier in the Arctic who is trying to change a flat tire might not be able to thread lug nuts onto the bolts. And when a vehicle is out of commission, the supply chain isn't moving as efficiently as it should.

Castellani is sitting in the lab on the third floor of the blocky academic building that houses one of the institute's ancillary facilities. Here twin chambers are connected to a state-of-the-art cooling system and wind turbine that circulates cold air through mesh metal walls. A team of privates with Velcro name tags and baggy uniforms has set up a mock experiment that Castellani says he can't actually let me participate in. Apparently I must have a note from my doctor. Having traveled halfway across the country with a plan to get submerged in Ranger-debilitating waters, or rained on for hours while wearing a rectal probe, I am a little disappointed when he says that I'll need to come back on a future date. Apparently the Army doesn't want anyone without the proper paperwork to get hurt on its watch.

After brushing my disappointment aside, I follow him into a chamber that is packed with a treadmill, a folding chair, and a card table. Castellani dons a black balaclava and a warm winter coat and takes a seat. I'm seated across from him in a thin dress shirt. The constant breeze through the vent is refreshing at first, but I can see how being in here for a long time would wear someone down. A soldier assisting Castellani shivers and clenches her bright white hands. Castellani nods in her direction and says that she must have Raynaud's disease, a condition that affects mostly women and that makes them particularly sensitive to the cold. Meanwhile Castellani looks positively cozy in his getup. He waves the soldier forward and she deposits a foot-and-a-half-long pegboard in front of him. It looks like some sort of child's game from the 1950s, with two parallel rows of round holes, two collections of short metal pegs, and thick ring washers.

The Purdue Pegboard Test is a standard assessment of manual dexterity. The goal is to figure out how long it takes a person to grab the small metal pieces and insert them in order down along the rows while also fitting the smaller washers over the standing pegs. It's not exactly a fun game; the minute movements look annoyingly difficult to perform. I give him an incredulous look—this is the test I needed a doctor's clearance for? Furthermore, I wonder if it is really an objective way to measure how cold might stop a soldier from, say, changing that tire in the Arctic or tapping a few buttons on an iPhone?

Just about everyone knows what it is like to have their fingers slow down in the cold, but in a military setting these physical losses can have serious consequences. In battle this could mean difficulty reloading a weapon or working the dials on a radio. At some point, Castellani says, he'd like to try a different standard that might be a little more applicable to life on the front lines, and have soldiers time themselves field-stripping their weapons at various temperatures. The only catch is that he is having "a little trouble getting permission to have a weapon in the lab." It's a nod to a byzantine military bureaucracy that allows him to submerge soldiers in ice water but not give him access to the very rifles they will actually use. So the board game will have to do. He shrugs and then fits an inch-long peg into a hole for me. I am not exactly awed at the scientific majesty of it all.

Still, all of this is really just the precursor to the real goal of the study. Castellani's mustache peeks over the rim of the balaclava as he explains how the military thinks it might be able to trick the body into achieving a little more dexterity. For the first time in the day, I realize that the Army is working on the wedge, too.

Castellani tugs the bottom half of his balaclava over his head and runs his finger down from a spot just above his temple and along his cheek. He traces the path of what he says is the trigeminal nerve. "It turns out that this nerve is responsible for a lot of the reactions that your body has to the cold, and we figure that if we can make it think it's warmer outside than it is, then the body might produce a less severe vasoconstriction response," he says excitedly. More blood in the fingers means more dexterity. And more dexterity means that soldiers should be able to change car tires and fieldstrip their weapons faster, too.

The idea originates from a quirk in human anatomy. The trigeminal nerve is particularly well-positioned for body hacking. In medical diagrams it looks something like a chicken's foot with three fingers fanning out across the face. Yet the nerve's root cuts back into the skull and connects directly into the thalamus—the brain structure that controls temperature regulation—without any intermediary filters. Most peripheral nerves in the extremities take a much more circuitous route to the brain. In effect, the sensations from this nerve go to the brain quicker than other ones do, and Castellani believes that it could be a sort of thermal shortcut that just *might* have an effect on the rest of the body. His hypothesis is that if he can keep the trigeminal nerve toasty, then vasoconstriction might be less severe around the entire body, and perhaps his soldiers can get just a little more performance out of their fingers. If that doesn't work, he's also thinking about inventing some sort of heating patch for soldiers' forearms so as to trick nerves closer to the source to restore blood flow—because hell, if body hacking doesn't work, there's always the possibility of just inventing better gear. Only time will tell whether these tests will bear fruit.

Results from the dexterity testing might come out in a year or two, but there is at least one training protocol that USARIEM developed that is already standard operating procedure for soldiers fighting abroad. Since September 11, 2001, the military's main areas of operation have been in the high ranges of Afghanistan and in the scorching deserts of Iraq. The US Army has more than 150,000 soldiers deployed abroad, and most of them have to rapidly adjust to an entirely new climate and be effective on the ground almost from day one. While altitude tables might predict how many soldiers would fall out of missions in Afghanistan, the Army had no clear way to quickly acclimatize an entire fighting force to manage the Iraqi heat. Acclimatization is a process that involves a host of bodily changes that optimize heat dissipation. Sweat glands, plasma volume, skin, circulatory system, and metabolic rate all play a role. Until the necessary changes occur, a human body is extremely susceptible to potentially fatal heat stroke.

It turns out that just enduring the weather isn't the quickest way to change the body so that it can withstand a new environment. Rather, the fastest way is by working hard in it—a process that is strikingly similar to the work being done both by November Project and Brian MacKenzie.

Castellani helped discover that moderate stints of exercise in high heat for just a few hours a day before traveling can jump-start the fundamental acclimatization changes. "In the Army we ship people down to Arizona or the California desert for two weeks before deployment and have them running around outdoors for at least two hours a day," he says. Indeed, research dating back almost 20 years shows that high-intensity training in hot conditions for just 5 to 8 days will greatly increase a person's ability to survive in the heat long-term. The stress from the workouts is a sort of wedge that speeds an array of physiological changes.

But the benefits don't stop there. There's some evidence that heat acclimatization might even help a soldier survive a bomb detonation or bullet wound. Recent research on animal subjects out of Israel show that heat-acclimatized mice survive traumatic brain injuries better than their temperate-weather-dwelling compatriots. In that study, scientists fixed mice onto a board and fashioned a mini-guillotine device that, instead of a blade, dropped a 95-gram weight on a targeted spot of their brain's left hemisphere. The mice were then given 42 days to recover before the researchers finished what they started and dissected the mouse brains to see how they healed. Surprisingly, heat-acclimatized mouse brains mended up nicely, while mice that never had to adjust to a new climate still displayed gaping wounds in their gray matter. Even though the authors of the research focused on specific neurotransmitters that fire during acclimatization (the research was for potential pharmaceutical development), John Castellani sees it as a tantalizing confirmation that getting soldiers into heat-fighting shape before they set on Iraqi or Afghani sand can help them heal better from wounds they might get during deployment.

There will probably never be a time when we will know enough about human performance in the cold that it actually makes sense to do battle in the depths of a Russian winter. Yet military triumphs and losses have long formed the test cases for probing the core of our shared human mettle. No one is ever more alive than the soldier whose life hangs in the balance. Time and again, generals have asked more from the bodies of their soldiers than was possible for them to give. The results, while occasionally heroic, are also far too frequently tragic. The shot of endorphins that the nervous system gets while struggling against the long odds of a military engagement could work as a sort of wedge into deeper levels of our mind–body connection. But there are other—safer—ways to get deep into the vital principle that makes us human. And that's why I'm headed to England.

12

TOUGH GUY

A MESOPOTAMIAN GOD in a sparkling gold Speedo chortles over the body of his recently vanquished opponent inside a wrestling ring located in a space somewhere in Central London that sometimes doubles as a fetish club. Blood streams out of Mahdi Malik's nose and pours out of his mouth. For a match in which his victory was preordained, it very nearly went the other way. In one high-flying assault a few minutes earlier, El Nordico's boot made contact with Malik's face and sent him over the ropes. He was unconscious when he hit the ground. It took almost 30 seconds for him to rouse and finally stagger to his feet. There was no way to know if something was broken. Never mind that, though; the show must go on. Eventually Malik mounted the ropes one last time for a final turnbuckle leap to victory. Now he's yelling insults at an audience of 30-somethings while two extremely ripped ring girls in otherwise identical red-and-blue latex body suits and face masks cavort behind him.

"I am your god!" he bellows. "You will all kneel before me." A few drunk men in the audience fire back that his Speedo leaves very little to the imagination and that none of them are particularly impressed. The scene vamps into a string of wrestling-inspired skits, burlesque performances, and, eventually, naked fire dancing. Once he's been dragged off stage, El Nordico, aka Ed Gamester—a heavily muscled and tattooed man who resembles a Viking even when he's not wearing makeup—saddles up beside me, places his heavily tattooed arm over my shoulder, and smiles widely.

"*This* is how we get ready for Tough Guy," he says. Ed is a self-described swashbuckler and the leader of Ghost Squad, a team of wild men in face paint who patrol the Tough Guy course and rescue people who get in trouble. It's my first night in London. I got off a plane just a few hours earlier and Ed is my host for the next few days as I compete in the oldest, and possibly most difficult, obstacle course race in the world. Long before Tough Mudder and the Spartan Race began to clutter up Facebook feeds—indeed, almost 20 years before Facebook even existed—Tough Guy sent tens of thousands of people through military-inspired boot camp hell in the most frigid weather England could offer up. The course is a series of soaring rope obstacles, sewer pipes, barbed wires, and, most gruelingly, ice water-filled trenches designed to get people wet early and then keep them cold throughout the race. Traditionally held in the last week of January, the course a few years ago in 2013 was so covered in snow and ice that they almost had to cancel it. Someone made a call to keep it up and running, and as a result more than 300 people ended up in the hospital with hypothermia. Tough Guy is a subzero sufferfest, and it's just the sort of challenge that I want to try to run shirtless.

There is something deeply satisfying about hearing English accents spout from underneath Mexican luchador masks, and I'm told that the club here is the only one of its kind in the United Kingdom. Also with us is Scott Keneally, the 6-foot-5 documentary filmmaker who introduced me to Laird Hamilton. Scott has long hair and, like the growing number of people I've been meeting at these kinds of events, is covered in webs of tattoos. He also has a contagious smile that's surprisingly effective at convincing friends to do stupid things with him. For the last 3 years he's been working on a movie called *Rise of the Sufferfests* about the sudden explosion of obstacle course racing (OCR), which has become one of the most popular weekend sports in the world. His reporting, which has appeared in a series of magazine articles for *Outside* and *Men's Journal*, traces injury rates and controversies in the OCR business. He's just flown in from Northern California, and we're both here to run Tough Guy. But for him the trip has a more personal project at stake: He's going to be screening his film for the first time in front of a small audience of racers. Among them will be Mr. Mouse, the octogenarian grandfather of all obstacle courses, who may or may not like his portrayal in the film.

Mr. Mouse, aka Billy Wilson, is a former member of the British Army's Grenadier Guards and carried a light machine gun in their Cyprus and Suez Canal campaigns. Though he saw some combat he spent most of his tours as the regimental barber, shaving men's heads and grooming their mustaches. When he retired from service he moved to a rustic, sprawling farm in a small village outside the town of Wolverhampton in the English midlands. The house was once briefly the home of William Shakespeare, or so he claims. Wilson started a string of successful barbershops in the town, and he made a name for himself with outrageous promotions featuring topless women he said were his staff barbers. Customers queued up for cuts and took their spot on his chair even though the buxom blondes were pure fiction. He made a fortune, but it wasn't enough for him. He wanted to make a mark on the world. An accomplished runner, Wilson believed that it was possible to help move toward world peace through sporting events. If he could find a way to direct the violent tendencies of wayward youths toward sport, then he might be able to mold them into productive members of society.

This is why Tough Guy is more than just a race. It's the spiritual center of the entire obstacle course race industry. In much of the modern world there is no clear line between what makes a boy and what makes a man. In bygone eras youths might head off to war or kill a lion to prove their mettle. The dearth of those sorts of experience—not to mention the societal costs of large scale violence as a way of proving some vital aspect of self-identity—leave many people wondering what they are capable of.

Tough Guy was a synthesis of Mr. Mouse's military background and a sports initiative that he started after the first London Marathon in 1981. Staged on the grazing fields behind his house, he dug long trenches in mud and filled them with water. "Mr. Mouse's Home for Unfortunates," as the farmstead came to be known, turned into a sort of runner's carnival. There were hay bale jumps and a few small wooden obstacles, but it was a pretty barebones affair by today's standards. Still, the real challenge wasn't the physical exertion; it was the cold. Planned for the last week in January—the week that tends to be the most frigid in the UK—he wanted competitors to power through snow and ice. He wanted them battered and frozen to a point where they could almost feel the fingers of death. Only then, in the face of their own mortality, he said, would someone know who they really were. Even so, since its inception there has been only one death on the course. In 2000 a runner named Michael Green collapsed on the course and died of a hypothermia-induced heart attack. Injuries are another matter—there have been no shortage of broken femurs and pelvises at local hospitals on race day.

The event was an instant hit, attracting thousand of contenders from across the country. As Tough Guy grew, Wilson added obstacles. Wooden towers with cargo nets soared four stories out of the muddy ground. He strung barbed wire over icy pits, installed sewer pipes for people to climb through on their hands and knees. In one particularly vicious section he hung electrical wires charged with tens of thousands of volts of current. It was brutal. And for close to 25 years it was the only event of its kind.

That's when a young Harvard MBA student named Will Dean got in contact with Wilson, inquiring if his brand of torturous races might have legs in the United States. Dean, a doughy Englishman with a shapeless head of hair, claimed to have worked as a counterterrorism operative for the Special Air Services (SAS), a storied British special forces unit, before joining the business world. According to a lawsuit, Mr. Mouse alleged that over the course of several months Dean wooed Wilson with the prospect of a global franchise of obstacle course races. Dean took careful notes and copied obstacles and marketing materials in service of a plan to build similar obstacles in the United States. Then Dean simply stopped contacting Mr. Mouse.

Not long after, at a ski area near Allentown, Pennsylvania, Dean launched Tough Mudder. He promoted it with $8,000 in Facebook ads. The event was an out-of-the-box hit. Tough Mudder started expanding almost before the first race was over. To this day Wilson claims that Tough Mudder simply knocked off his brand. There were years of bad blood, and eventually an out-of-court settlement in which Dean gave Wilson $750,000. Years later there's still enough animosity to make dropping Dean's name in Wilson's company a dicey affair.

Simply bringing up Dean works Wilson into a frenzy in which he starts to question every aspect of Dean's resume. He claims that there is no proof Dean worked for the SAS, as the military rolls are kept confidential.

Whatever the particulars, one thing is certain. Dean grew Tough Mudder into a global brand in a way that Tough Guy, with its almost 30-year-long history, never did. Now there are spinoff franchises, TV shows, sponsorship deals, and millions and millions of grimacing social media posts. Mr. Mouse has been left behind. While he might still run the most brutal obstacle course race in the world, any way you look at it, without Will Dean, OCRs might never have gone mainstream.

Back at the London fetish club turned wrestling arena, Scott Keneally stands in line for yet another beer from the bar. By his own estimate he's been drinking pretty much nonstop since catching the flight from San Francisco, ordering four miniature bottles of wine from different flight attendants and then, once landing, finding ample opportunities to slip into one bar after another. As a naked fire dancer twirls two torches on chains in the middle of the ring and then drags the cat-tail-looking lit end across her breasts, Keneally confides in me why he was nervous about how Mr. Mouse might assess his role in his film. From a certain standpoint Mr. Mouse is the hero of the businesses. He's the father and the heart and soul of it all. But he's also its biggest loser. "I don't want to break his spirit," Keneally says.

The screening will be held after the race. If it doesn't go well, it will be Keneally's last chance to run Tough Guy. He's sure that if Mr. Mouse takes affront to his portrayal he'll be banned from taking part again. For now, however, there seems to be only one way to get ready for the event—by drowning any anxiety with whiskey, rum, and beer. The night continues from ringside to a hazy ride on the Underground and ultimately to Ed's flat on the outskirts of London, where we eventually stagger into bed at around 4 in the morning. Before we fall asleep, Scott cries out "Shotgun!" in an attempt to reserve the front seat for tomorrow's drive. To the surprise of no one, we don't make our 9 a.m. wakeup call. But we do eventually pile into a car by noon. Ed packs it with all of the props he'll need to charge across the course as the leader of Ghost Squad: several tubes of body paint, armor, a few swords, and maybe a half-gallon of paraffin lamp oil—the key ingredient for fire breathing.

I cram into the backseat next to Ed's brother, Will. He is a long-haired, thinner version of Ed, and he bears a passing resemblance to the character Inigo Montoya from the *Princess Bride*. When Will tells me he'd never seen the movie I immediately pull out my smartphone and bring up a YouTube clip from the movie. I tell him to repeat one of the most memorable lines from the film: "Hello, my name is Inigo Montoya. You killed my father. Prepare to die." He does, and while the impression needs work, it's still satisfying. However the next few hours on the road into the English midlands seem to drag on forever. Ed bobs and weaves his tiny car through traffic and through his own pulsing hangover as Keneally curates a mostly hard rock playlist on the stereo. Along the way they school me on the idiosyncratic biography of Mr.

Mouse, making several caveats at improbable and contradictory plot points by saying, "Well, no one is entirely sure how much of what he says is truth and how much he just makes up." But their tone also bears a cadence of reverence and loyalty to the man they call the "Madman of the Midlands."

"The first thing you will notice when you meet him is that the house smells like the sixty or so dogs that he keeps there," says Ed. To which Scott corrects him, saying, "It's more like you will be walking inside of a dog's lung: moist, heavy, and absolutely overpowering." It's not long until I am smelling what he's describing.

After several wrong turns prompted by the failings of the GPS, we turn past a bright neon sign announcing "Registration." It's two days before the race, but the farm is buzzing with activity. A massive, coal-fired fireplace sits at the very center of the Home for Unfortunates. It's an odd construction made out of four interlocked shovel scoops that have been salvaged from farmyard backhoes. An ash heap at least 3 feet high rises out of its center, and the residents and workers at Tough Guy use the conflagration to dispose of just about everything—plastic bags, beer bottles, stray office papers, and unfinished meals. Much of the pit's foul-smelling smoke makes its way up an ancient stone chimney, but enough escapes into the house in general that carbon monoxide poisoning seems like a real possibility for anyone who spends too long indoors.

A failed piece of taxidermy sits in a back corner of the house: a full-size pony—one of Mr. Mouse's dearly departed pets—that is covered in a centimeter-thick layer of dust and debris. Sitting across from that is a fully operational Bren machine gun from Mr. Mouse's military career. The weapon is mounted on a tripod and points into the middle of the room. It's probably not loaded, though the haphazardness of the scene makes me hesitant to try the trigger.

Despite the chaos, this is also the nerve center for the upcoming race, a place where insiders to Mr. Mouse's athletic fiefdom come to warm up, share gossip, and reminisce about their accomplishments on the field. A pile of pasties—or Cornish meat pies full of ground beef and potatoes—are free for the taking at the center of the massive banquet table along with a seemingly endless supply of beer and wine.

In the group is Clive Lange, a seven-time Tough Guy veteran. He stands more than 6 feet tall, with hair closely cropped around a receding hairline, and possesses the muscular physique of a man whose day job is personal training, sports massage, and karate. He spends the first few minutes of our conversation describing how much pain he was in during the previous year's race. Apparently, the sharp skin of ice on the top of the obstacles cut through the skin on his shins and left him bleeding. Most past participants rate the experience akin to torture, but they also can't stop reliving the glory of their accomplishment. "There's nothing like this out there. When you finish you are broken and bleeding, shivering and exhausted, and never want to do it again. But then, five minutes later when you are warming up by the fire, the endorphins start flowing and all that goes away. It's like a drug and you know you're going to come back," Lange says, likening the event to what he imagines childbirth must be like. His metaphor concerns the moment when a woman on the verge of giving birth curses her husband's very existence for his role in conception and then, when the baby is in her arms and the pain is meaningless, only the child matters. Despite his fondness for counting Tough Guys as children, Lange will sit out the race this year. As a member of Ghost Squad, he will instead cover himself in war paint and cheer on the people brave enough to line up at the starting gate.

Also sitting out this year is James Appleton, a three-time winner of the event. When I reach out to shake his right hand he waves me off and offers his left one instead, leading to a moment of confusion. His right is broken, the result of an unfortunate injury resulting from a night of New Year's revelry. He's not happy about it, and as one of very few champions in the world of obstacle course racing he knows that as the sport grows there's at least a chance that it might spawn a professional category that could provide him with a living wage. For now, though, Appleton covers the bills by working as a travel photographer shooting landscapes and weddings around the world that often happen to coincide with the start dates of different races. In 2014 he achieved a moment of international fame when he finished third in Tough Guy but was so broken and incoherent at the end that he could have died of hypothermia. The winner of the race, Norwegian runner John Albion, helped him into a shower in a viral video captured by Scott Keneally's film crew. In it Appleton shakes so severely that it seems like the convulsions come from the very core of his being. Even a layer of neoprene couldn't help him. In the video he slurs his words as he attempts to explain how he feels: "After the underwater tunnels I lost all sense of time and didn't know really what was going on . . ." he says before trailing off. In the later stages of hypothermia a person loses his ability to think. He could have died, but luckily it only took him 4 hours of shivering by the fire to regain control of his body.

This exact sort of brush with death is the point of Tough Guy.

The race isn't only about overcoming barriers; it's about emerging a totally different person. "At the start line I remember how terrible it was the year before and I never want to do it. Then, before the cannon goes off, I know I've made a choice and that there is just nowhere else to go but through, and I get this feeling like I'm Ironman putting on my armor. The armor comes on and everything just tightens and there is an impenetrable seal on my heart," Appleton tells me. That seal obviously has limits, one that nature can inevitably surpass and crack. But it's also a seal that lets him manage sub-5-minute miles through knee-deep water, over hills, tunnels, barbed wire, and shoe-sucking mud. It's a seal that I think I might be able to conjure up the very next morning without a wetsuit, without compression gear, without really much at all.

The wind isn't as bad as it could be as I line up on the starting line the next day. My race number puts me at the front of the pack. I'm mostly naked from the waist up, clad only in a pair of blue running shorts, gloves, and a bright orange hat. On my feet are a pair of worn-out running flats that are as close to barefoot as a sneaker can get. A large berm surrounds the racers on two sides with a crowd of photographers and voyeurs cheering and pointing. Minutes earlier, Ed Gamester—dressed like a Viking—smeared a handful of red paint on my chest and rendered a Celtic rune on my back. "It is the rune of victory in single combat," he growls before taking his place at an 80-gallon steel drum and starts pounding away with a pair of broken chair legs. The cacophony of excitement builds as I shuffle my legs at the starting line and take a deep breath to stay warm. Other people are rubbing their arms for whatever fricative heat they might generate. Standing around for the cannon blast is maybe the worst part. A shiver threatens to move down my spine, and I try to force it away. And then I feel it, the same tightening that comes when I dip into ice water or turn a hot shower to cold. The switch, whatever it is, flips on. The cold will not penetrate.

Someone launches a pair of purple smoke grenades out onto the field and as the cloud billows into the sky the cannon goes off and a thousand or more people charge forward with a war cry fit for the Highlands of Scotland. The 25 main obstacles on the course are rated by the designers for both their fear and pain potential, with the bulk of the most formidable challenges saved for the last few miles of the race. The initial sprint out of the gate is a fairly straightforward country run through a muddy field that is interspersed with only a few raised log hurdles. We pounce over them. The wind hasn't picked up yet and everyone is in good spirits until the first water obstacle—a waist-deep plunge into muddy January water.

For most runners, this is the moment when Tough Guy changes from an ordinary obstacle course race into an impending brush with death. While the cold shock isn't particularly onerous on its own, every runner realizes that it's the first hit to their core temperature, and that they will be wet for the rest of the 10-plus miles to come. Their shoes will be soggy and sucking with every step, their limbs slowly growing stiff and numb.

After 6 months of training, when I launch into the water pit my initial reaction isn't panic or pain. I feel instead an overwhelming sense of joy. The months of cold exposure that I've done leading up to this moment unleash a giant wave of endorphins. But that joy is tempered when I realize that my challenge for the rest of the race isn't going to be conserving heat; rather it's going to be about the ordinary physical exertion that comes with running up and down hills and traversing hand-over-hand rope courses. One thing that I've learned about maintaining my body heat is to simply think warm thoughts while in the cold. Sometimes I'll conjure images of a campfire in my belly, but today I decide to imagine there is a fire-breathing dragon in my gut. This creature from the bowels of hell is going to keep me warm, so help me god. My breathing sounds like a growl as I mimic the sound of fire. It's a silly image, almost childish. But it works. My lips curl into an invincible smile that course photographers seem to capture with just about every click.

My pace is far from game changing. I never had hope of winning Tough Guy, nor do I even consider myself fit enough to keep a place at the front of the pack, but as I look around at the other competitors it's clear that my relationship to the wind and water is somehow different than the other contenders. After a brutal series of hills I eventually catch up to Scott Keneally, whose "Suffer Club" T-shirt is soaked through against his skin. He's shivering a little bit. I realize that there's an added advantage to being shirtless—I'm not carrying any water with me as I run. After every obstacle I have a chance to dry off where nearly everyone else stays wet.

Just as we get to a series of water-filled pits termed "foxholes," Keneally tells me that he can't feel his toes anymore. Every obstacle we jump into is filled with waist- or chest-deep water, and we haul up a steep muddy slope on the other side. In most cases the slopes are so high and slick that it's hard to find any grip or purchase to leverage yourself up and over. So racers offer helping hands to one another. It's a sort of a factory-like procession: Let one person help drag you up and over one mud slope, then turn around and assist the person behind you. We emerge covered in a thick layer of mud, soaked to the bone. This is also the single most gratifying moment in the race. I quickly realize that I'm not just trying to push past my own boundaries but that somehow we're all here struggling in the mud together.

In this moment of communal suffering I'm part of something much greater than myself. I'm no longer a competitor, but a single cell in a giant struggling body of humanity. The course breaks down the barriers between people so that after I help push a struggling woman up one wall it is perfectly normal to accept her hand once she's at the top.

Scott Keneally stays behind at the pits. He is clearly enjoying the moment of community, helping a hundred or more people up. He doesn't worry that his time to complete the race will likely be more than 3 hours. A few months earlier he completed an event called the World's Toughest Mudder—running 50 miles of terror-inducing obstacles in 24 hours. He doesn't need to prove himself anymore. He just wants to bask in the moment.

After the foxholes, the frequency of obstacles picks up, making the last few miles of the race far more difficult than the start. There are hanging ribbons of electrical wires carrying 10,000 volts of muscle-seizing electricity, and a three-story woodworks called "Goliath" with cargo nets hanging off it. I run to the base of the monstrosity and look up to see Ed's brother, Will. He is covered in white and black paint, pointing at me and urging me up. Will shouts down at me that I'm a madman for going shirtless. I grin back my happy retort, "I'm a fucking dragon." Fire burns in my belly.

I jump onto the net and push up with my legs, taking two rope rungs at a time. The net bounces with the weight of 10 people all climbing in unison, and as I crawl over the top I hear a scream behind me followed by a loud thud. Someone calls out for a medic and I look down to see a 20-something woman lying on the ground, dazed. Somehow she fell through the cargo net and missed a mesh safety net below, falling at least 30 feet onto the muddy earth.

At first she can't move, but soon she staggers to her feet and begins to run again, charging up a second cargo net as if nothing happened. The only sign of the near catastrophic injury is a 6-inch long rip in the seat of her black yoga pants. We run apace for a while until I'm sure that she's okay, and then I press onward to an obstacle called the underwater cavern, which is inspired by the tunnels where soldiers chased the Viet Cong. It's a site that Mr. Mouse warned me about earlier that day, where the electric shocks caused one competitor to have a heart attack and then pass out face down in standing water. The man was resuscitated a little while later at a local hospital. Mr. Mouse subsequently drained the water from this part of the course to stave off future deaths. So I move past the wires and enter a concrete sewer pipe that is too narrow to allow me to crawl. Instead I slither along the muddy pipe, inch-worming on my back in total darkness until I see a circle of daylight where the tunnel curves sharply upward.

I emerge with a newfound respect for claustrophobia. I must have taken longer in the tunnel than I thought, because Keneally is right behind me. He comes next to me and says that the next challenge is the one he has nightmares about in the year between races. It's an icy pond of water at least 5 feet deep. Five logs form a sort of bridge across the middle, and racers have to dunk their heads beneath the wood and swim along the expanse. Keneally is already freezing and he hesitates when the bridge comes into view. Still warm, I jump into the water and take a few strokes. It must be barely over 30 degrees, but I'm still smiling, somehow still elated after hours of cold running. We paddle to the logs and Keneally goes first, groaning audibly every time he comes up for air. James Appleton has spotted the two of us, and is taking pictures of the event. He captures our faces as we emerge like the undead.

Going completely underwater is a different sort of experience than submerging oneself just to the neck. The nerves in the head are closer to the brain, and the cold water courses over the trigeminal nerve (recall that this is the nerve Castellani is using to hack soldier physiology), meaning that the cold signal to the central nervous system is stronger.

On the other side of the bridge Keneally presses his hands to his temples in an effort to soothe what is effectively a massive ice cream headache. When I dunk down under the water to meet him I also come up in pain. Time seems to slow down and I lose my balance, stumbling onto my left side. Appleton snaps a series of photos as I regain control. Twenty minutes later—after barbed wire, a fire jump, and another stretch of icy water, Keneally and I jog to the finish line.

I forget to check my official time for the race, but someone records it in a notebook at the line, taking the number off of the muddied bib I pinned to my shorts that morning. Champions like James Appleton finish the race in an hour and a half. I guess that I'm closer to three or three and a half. However, where Appleton almost collapsed from hypothermia last year, I'm still on fire, standing shirtless at the end of the course with my chest heaving from the exertion, not the cold. At some point, as my body tries to warm up, I expect that I'll have afterdrop when warm blood flows through my cold limbs once vasoconstriction relaxes. But 20 minutes later, even when I return to the fire at the center of Mr. Mouse's home, it never comes. I warm up comfortably; victorious.

The course remains open for another few hours as people take their time finishing the race. In that time Ed and James come up to me asking to learn the method. They say that I was the only man smiling throughout the whole race. I might not have been the fastest—really I wasn't even close to it—but I was the happiest. "If I can take a piece of that with me once my hand heals, who knows what I'll be able to do next year," Appleton says.

The next morning Scott Keneally screens his movie for Mr. Mouse and a few select runners and course volunteers. Keneally watches his mentor closely as he breathes heavily during the descriptions of his own madness and the way that the industry and its fortunes were stolen from him. It's not the final cut. The sound isn't quite right just yet, and Keneally's voiceover is a muddy recording from his iPhone. But when Mr.

Mouse smiles as the credits roll, Keneally knows that he's captured something of Tough Guy's spirit.

For me, however, Tough Guy is really just the beginning, a modest test of a method I think has unleashed the dragon in my belly. The real battle isn't against mud obstacles or ice water, where paramedics or the Ghost Squad are just a shout away. Now my mind is set on conquering a mountain that claims a fistful of lives every year.

13

KILIMANJARO

THE WORD ON the street in Moshi, a dust bowl of a city about an hour away from Mount Kilimanjaro, is that some cell phones will work near the summit of the mountain. Not one to miss the chance to post a shirtless selfie from atop the imposing peak, I set out to locate anyone who can sell me an off-market Vodacom SIM card that might work up there. Just outside the bus station a dozen or more men sit behind fold-up tables with an array of cracked and scratched but otherwise indestructible Nokia phones. It's a humble front for a technological revolution in this part of Africa. In Tanzania, cell phones are the lifeblood of the online community and most people's primary access point to the World Wide Web. Absent a reliable banking infrastructure, swapping phone credits is a proxy for banking, and mere ownership of a phone is an analogous status symbol to having a burnished aluminum MacBook in the United States. I locate one suitable salesman, a man in his twenties with a gold tooth twinkling in his grin, and trade him a fistful of bills numbering in the tens of thousands of Tanzanian shillings for a SIM card that is a centimeter too large for my iPhone's tray. To make it fit, he grinds the plastic against a concrete wall until it approximates a more suitable shape. While engaging in the task he correctly notes that I—the 6-foot-tall white guy wearing a neon green hiking backpack on the streets of this otherwise unremarkable African city—am probably planning to make my way up the mountain. Then he asks me the most important question of the day: Which route am I going to take?

He, like most of the able-bodied men in the city, moonlights as a porter for foreign mountaineers. If there's room in the crew, he says he wouldn't mind a job. Most porters have charged up the mountain scores, if not hundreds of times and, besides cell phones, it's the biggest business in town. There are six standard routes up the mountain, arriving from most of the directions on the compass. The main routes—Lemosho, Machame, Marangu, Rongai, Shira, and Umbwe—are sometimes affectionately nicknamed after beverages. The most appealing of them is Whiskey, but we're going up Coca-Cola (Marangu), which, despite its soft drink moniker, has one of the lowest success rates of all the paths to the top. It's the same route that Hof has taken multiple times.

I mention the soda to the salesman and he lights up.

"So you will take six or eight days?"

"Only two."

He smiles a wide, toothy grin and gives me a tentative fist bump.

"You're crazy," he says, adding for good measure that he doesn't know anyone who would tackle the mountain in such a short amount of time. The conversation, which was leading him to getting a job as a porter on the expedition, peters out. It's not just that a short trip would pay less (though for the record we are paying our porters for a full expedition plus a fat tip). No. Huffing up the mountain so quickly would be too dangerous. Instead he offers up that he knows where I can get a quick prescription of Diamox, but I tell him that I have no plans to bring along any medication. Without any obvious ways of obtaining more Tanzanian shillings from me, he offers a bit of advice.

"Pole, pole" (pronounced pole-eh, pole-eh), he says, which in Swahili means "slowly, slowly." It's a phrase that I'll hear on the mountain almost constantly. Every passing porter, guide, and tourist greets each other with the same reminder: move too quickly and you won't make it. There's no need to rush up to the top. The risk of getting altitude sickness is very real.

After some jury-rigging, the SIM card fits into the iPhone's tray well enough to lock inside. To my surprise, the phone fires up on the foreign network and is ready to go. I hop a motorcycle taxi back to the hotel where the rest of the group is staying.

This is Hof's third group expedition up Africa's tallest peak, but will be the fifth time that he has actually climbed the mountain. It's been a record of unheralded triumph. In the previous 2 years the company managed a staggering 92 percent success rate to the summit. This is astonishing given that several people on the expeditions had been drawn to Wim Hof and his method because they were seeking relief from a debilitating physical ailment. The arthritic blacksmith Henk van den Bergh, for example, reached the summit the year before in just a shirt and shorts despite almost being immobilized by his condition just a few years earlier. The treks ended with shirtless group photos next to the highest point of elevation, and viral posts on social media circulated around the web until the next climbing season. For the most ardent followers of all things Wim Hof, the Kilimanjaro expedition is a sort of Holy Grail of human grit. As the most recent person to secure a spot on the climb I'm still catching up on names, bios, and reasons for coming along in the first place, but one thing about this expedition strikes me as different from what I'd heard about prior trips.

While everyone certainly has a dedicated regimen of ice baths and morning breathing routines, this year's crew of 29 people exudes a different, perhaps less urgent, feeling than the groups that came before. They are mostly Dutch—with just a few Belgians thrown in for good measure—and a surprising number come from the halls of big business. Three are managers from the Dutch banking conglomerate ABN Amro who fell into Hof's orbit because they felt that his techniques might help them become better leaders in their industry. There are also a smattering of Ironmen and CrossFitters, and one of them, named Dennis Bernaerts, hangs on Hof's every word like he is following a prophet. There's Bart Pronk, who maintains the 10-week online course on the method. Bart was by Hof's side all the way to the top on the last two trips. There's a barrel-chested 76-year-old man named Frances, and a practitioner of holistic healing who is currently fighting off chronic Lyme disease. One woman is recovering from lymphoma. Then there's Sylvia, the owner of a series of marijuana shops outside of Amsterdam who says she hasn't had a full night's sleep in more than 20 years. All in all, good people. The group is so hospitable to the lone non-Dutch speaker that the entire expedition decides that they'll speak English for the duration of the trip, a language that, owing to the fantastic educational systems of the Netherlands and Belgium, all the participants are fluent in. Even the gender mix is more equal than in previous years, with eight women on board for this journey, where previous expeditions managed only one or two. Despite all this, it's hard to fight the sense now that Hof has proven his technique so many times that most of us don't have any fear of failure. Not one person expresses any hesitation about reaching the summit, and it reminds me of the most important warning that firewalkers get before they run over hot coals: The only people who get burned are the ones who aren't nervous.

Of course the most certain of us all is Hof himself. He's planning to wear the same outfit that he wore on the plane from Amsterdam all the way up the mountain: a bright blue pair of swimming trunks featuring a school of tropical fish, safety-cone orange sneakers, and a gray tank top. In the blue plastic garbage bag that stood in for his carry-on bag he carries one other item, a blanket. This, his mountaineering-outfit-slash-formal-occasion-uniform, is all that I will see him in for the entire week we are in Tanzania. Always looking for a way to set the upcoming achievement apart from earlier ones, Hof announces that together we will smash his previous record by a half-day and make it to the top of the mountain in just 30 hours. This, he tells us, would be a record for the fastest unacclimatized ascent of the mountain by a group. The information seems to pass over us without comment.[1] It is as if merely being with Hof is enough to ensure success.

An e-mail in my inbox, however, tells a different story. When I learned of the speedy plan—one that even surpassed what I had told the tout in Moshi—I dashed off a message to John Castellani, the Army doctor at USARIEM who I'd met a few months earlier. I asked him to give me a realistic estimate, based on his experience and research, of how our group might fare against the potential ravages of limited atmosphere. Castellani relayed my inquiry to a few experts at the base, and within a few hours he wrote back that 60 to 75 percent of us should expect to get acute mountain sickness (AMS). To put that in perspective, that would mean between 17 and 21 of us could begin the spiraling downward effects of limited oxygen—starting with headaches and dizziness, and potentially ending in a fatal situation. Seeing the number at first gave me a sense of pride, as if it were a challenge: *Of course we will beat that number.* But I'd be lying if I were to say that my confidence wasn't followed by a twinge of fear. *What if something does go wrong?*

The reason that Hof's method is so effective against altitude sickness is that constant and rapid breathing compensates for the lack of available oxygen. Since every step up the mountain means that each lungful of air is a little less rich, the solution is just to breathe more. As long as you never let up breathing consciously, AMS should never really be a problem. But what would we do if someone fell seriously ill? The previous year's group boasted a 92 percent success rate, which means that at least a few people were not okay on the mountain. The medical report from that ascent mentioned that one climber's lungs filled with fluid and that his heart started beating irregularly. Would we all be able to keep up our breathing routine if we also had to respond to an emergency?

---

1   It's hard to be sure what the fastest group ascent of Kilimanjaro actually is because those records do not seem to actually be recorded anywhere. However, the fastest solo ascent of Kilimanjaro was done in 6 hours and 42 minutes by Kilian Jornet, a Spanish mountaineer and ultrarunner who had previously acclimated his body to the height.

Might one person getting AMS make other climbers lose focus as they tried to help?

There would be guides, of course, for safety. One burly Tanzanian named Mike Nelson, who would lead the local support team, boasts of having been on the historic mission to recover George Mallory's body off of Everest in 1999. His small team of climbers found the Englishman's corpse on the ledge where he'd fallen to his death 75 years earlier. The oxygen on Everest is so sparse that despite their best efforts they couldn't mount the strength to actually transport his corpse back to base camp. Instead they looted his pocket watch, clothing, and gear for what would become a museum exhibit. They then covered his naked cadaver with a cairn of stones. Almost a dozen men like Mike would be there for us if we got into trouble, but their presence is no guarantee of safety. If someone gets into trouble they might not be nearby because we won't all be moving at the same pace.

For my part, I've begun to wonder why I'd come to Tanzania in the first place. After almost 4 years of following the method, part of the impetus for coming on the trip was simply inertia—like what happens when a couple gets engaged and then drifts toward a wedding date despite whatever second thoughts they might be having. Getting to the summit is in my mind somehow inevitable. It is my goal to challenge myself with something impressive, but the specter of what *might* happen on the mountain plays on repeat in my brain. At best I am an amateur mountaineer—an avid hiker, but not someone who heads up into the mountains every weekend.

The grounding concepts of the Wim Hof Method are simple enough: By routinely stimulating a stress response a person can take some modest control of their fight-or-flight reactions. A thousand cold showers have made me a warrior in my bathroom and even the snow, but no human has ever evolved to thrive at 18,000 feet. The goal of the method isn't to make the practitioner invincible to the elements. The hundreds of thousands of fallen soldiers in the Siberian winter prove that in the battle against nature, nature always wins. Exercising the stress response just allows a person to assert a measure of control when the environment gets challenging.

There's really not enough time to get to know everyone in the group, but I pair up with a curly-haired man in his fifties named Emile. We size each other up as likely candidates who would keep about the same pace. He is fit, but more important, he always seems to have a smile on his face. He's my buddy for the trip, and together we'll be responsible for checking on each other's health. We each have a copy of the Lake Louise Score criteria, the method for rating the severity of acute mountain sickness that Geert Buijze recommended. Developed near the high altitude lake in Canada that shares the name of my mother-in-law, it is really just a list of potential symptoms next to a corresponding self-rating. We promise each other that if either of us exhibit symptoms that accumulate to a diagnosis of mountain sickness we'll start down the mountain together.

There's a local news crew at the park gate waiting to interview Hof. The papers are curious about the Iceman's feats, and he is happy to ham it up for the camera. He explains that the expedition will break a record for the fastest group ascent to the top of the mountain. He then does a series of splits and headstands so the crew will have some B-roll (supplemental footage) to go along with their story. When they're done he comes up to us to give his own version of Henry V's speech at Agincourt.

"This trip is not about just testing out limits," he begins. "It's about leaving our egos behind and delving deep into our physiologies." He pauses for a moment and smiles at a pun forming in his inspirational ether. "There is no ego on this mountain. Just *we go*." I do everything I can not to bury my face in my hands to stop myself from guffawing at the cheesiness. Given the speed of our upcoming ascent, he says, the real challenge on the trip won't be the cold but in keeping the oxygen levels up in our blood. "So don't worry about being shirtless the whole time. If the cold gets in the way then put something on and remember to keep checking your O_2."

It turns out that checking blood oxygen levels is quite easy. Just about every person has packed with them a small digital monitor that uses infrared light to measure O_2 saturation in the blood. The devices fit on a finger and give readouts for heart rate and oxygen. A healthy reading measures about 97 percent saturation or above. During the hyperventilation and retention exercises I routinely performed at my home in Denver, I could usually bring my levels down as low as 50 percent while holding my breath—a number that would indicate a chronic lung problem if it weren't temporary. But the higher we go on the mountain the lower our O_2 saturation will be, so the plan is to keep checking our levels every half hour and then start the breathing faster and deeper if we drop below 90 percent. The human body's natural response to high altitude is to start breathing faster—again that's also how Diamox works; it simply increases passive respiration—so we will breathe quickly and continuously from the base of the mountain all the way to the summit in order to get ahead of any deficit that might develop.

We let the plan sink in, turn our backs to the gate, and then start up a long dirt path into a canopy of jungle. A large sign at the Marangu gate serves as a warning to anyone attempting the 18.6-mile route to be aware of the perils of cold and limited air, adding that all "hikers should be physically fit." We all more or less fit the bill. Shirtless, the group winds its way past the sign. Behind us are almost 60 porters carrying the various overnight baggage, food, and medical supplies that we might need on the trip. They walk with bags balanced on their head and sometimes two or three hiking backpacks on their shoulders. Most of the porters are in their early twenties, and it strikes me that while we are planning an expedition to the summit, no one in the group is nearly as hardcore as the men carrying our bags. Most of them do the expedition every few weeks, which means that most of the porters are already acclimatized to the change in altitude. But that doesn't make their effort any less impressive.

The first leg up, to a hut called Mandara, traverses a steep path through densely packed jungle that would look at home as a set in an Indiana Jones movie. Small black monkeys call to each other from the forest canopy and flightless birds hop through the underbrush and hanging mosses. Focusing on our breath, we are mostly silent on the climb, but we take time every now and then to admire some small piece of wildlife—a chameleon with pinprick eyes trying to blend into the surroundings or an iridescent white slug crossing the path. At first we try to stick closely together with Hof taking the lead next to Frances, the barrel-chested 76-year-old mountaineer. We move at a slow pace for about an hour before Hof seems to grow restless. He huffs audibly, sidesteps the older man, and pushes past him. In a few minutes he puts a hundred yards between himself and the group, and then he almost disappears into the brush so that we only catch glimpses of his bright orange shoes now and again.

The separation isn't for very long though, because within a few short hours we burst out of the rainforest to a small series of A-frame huts for a light lunch. The meal consists of a fried half-chicken that has soaked in oil for so long it resembles the undead. There's also a sampling of fruit, a dry dessert cake, and a tub of yogurt. The meal's mere size, even after factoring in its relative unpalatability, is an assurance that we definitely won't starve on the ascent. Just as we set our teeth to the rubbery chicken, the sky outside the hut opens up with a deluge of water. The rainforest, true to its name, is making itself known before we reach the next climactic zone.

Most people coming up the mountain would stay in these huts overnight so as to slowly acclimate their bodies to the altitude. Though the rise from the gate to here only brought us to 8,860 feet, recommended protocols make ascending the mountain a painfully slow process. I suspect that very few hikers feel the *need* for a pause at this juncture but stories of the sicknesses that afflict most climbers are rightfully warning enough. It's somewhat refreshing to me that we are going to pursue a more aggressive pace, as stopping at just one in the afternoon would make me go stir crazy. So we look out at the torrent of rain as it fills the path with bright muddy puddles and decide to forge ahead.

No one, of course, likes being cold and wet. I'm not sure exactly who among us first unzips their pack and ferrets out their rain gear, but when the first coat goes on the urge for protection against the elements seems to spread like a virus. For me, the hardest part of the cold exercises that I have been doing for four years is always the moment when I decide to actually get cold. Whether it's standing above the waters of a frigid lake, near a fresh pile of snow, or even just in a blazingly hot shower with my hand ready to turn off the hot tap so only cold water flows through the pipes: the decision point where I jump in, lie down, or turn the faucet to cold is when my mind throws up the most resistance. The anticipation of discomfort is almost always worse than the actual experience. The inverse turns out also to be true. If one person acquiesces to comfort it becomes that much easier for other people to follow suit.

So before we are even outside of the A-frame the group is layered up in suits of rubber and Gore-Tex. Hof dons the trash bag that was his carry-on as a makeshift poncho, and I throw on a bright yellow slicker. To me the clothing bears the taste of failure. Only a few hours in and I've already broken the skin-out covenant I'd promised myself I would keep. The troop marches forward up the hill, all breathing in unison. It doesn't take long before I start sweating like a man in a rubber suit working out in the tropics, which, of course, I am. A half mile from the cabin I'm almost as wet on the inside of the jacket as I am on the outside. So I decide to stash it in my pack and let the water fall on my skin as a welcome relief. A few other people follow suit, and there is a tenuous balance in the group between the people who seek protection and those who are trying their hardest to remain pure.

Soon the foliage of the rainforest starts to grow smaller and recede as we enter into a more arid zone. We're above the rain and the sun brightens an otherwise gray sky. Now boulders, small cacti, and bristly shrubs pop up on the landscape. Our views extend to the extinct lava cones that make up the foothills of the volcano itself. Rather than take it all in, however, I mostly just look at the shoes of the man in front of me: Stef van Winkle, a nearly 7 foot tall Dutchman who is also, incidentally, the youngest person on the trip.

He's breathing hard, harder than anyone else in the line. Every minute or so he starts a succession of 30 rapid breaths and then holds and twists up his face so that he turns slightly red. He's practicing a technique that Hof gave us for when we start experiencing altitude headaches. The method rushes oxygen to a specific part of the body and works equally well for sore limbs after a workout as it does for a headache. Like many things with the Wim Hof Method, it's something that is easier to feel and experience than it is to describe in words. Hyperventilation decreases CO_2 and increases overall oxygen in the bloodstream while moving blood pH from acidic to alkaline. During retention (with lungs full of air) you sequentially tighten all the muscles from your extremities toward the place where you want to move the blood supply. The process is a little bit like wringing out a wet rag from one end to another and pushing the water out. It's hard to say exactly what you're physically moving inside the brain when you're targeting a headache; perhaps it's the smooth muscles in its vasculature, or perhaps you're just thinking at a particular spot. In my experience, after a few seconds of using this method minor headaches vanish and major ones decrease their intensity.

Van Winkle's breathing patterns worry me, though. We're barely over a third of the way to the top. Whatever benefits the breathing might lend, it's a bad sign that he's already fighting for oxygen. I ask him how he's doing and he gives me a thumbs up with a nod. But his gesture doesn't hide the pain. So he keeps his head down and continues the slow march upward. I pass him and continue to move forward.

Seven hours into the march, the group is spread out like a segmented line of ants interspersed over what must be a few miles. My buddy is gone—so much for safety checks—and I think that I'm somewhere toward the front. The landscape is now dominated by some sort of large, spiny plant reminiscent of California yucca. I'm all alone save for one Tanzanian guide named Joseph who carries an overstuffed backpack. He has either found that my pace matches his, or he's keeping an eye on me. The sun sinks behind a bank of gray clouds and a light drizzle coats my skin, but I'm feeling great despite a sudden drop in the temperature. Joseph watches me from inside his puffy jacket, curious as to why I have not yet put anything on. We near an elevation of 11,000 feet. The air is much thinner than it was when we started, but the thinness also means there is less air resistance. Oddly this makes it easier to breathe quickly. My O_2 levels never seem to fall below 95 percent. Out of curiosity I ask Joseph to lend me his finger and let the device check him out as well. A second or two later it settles on 82 percent—not terrible, but far lower than I expect for someone who should already be acclimated to mountain ascents. We share a granola bar and he tells me that Horombo—the place where we will spend our first night—is not too far off.

So I double my efforts and push forward at a brisk pace. The promise of a warm sleeping bag on a hard cot might as well be spurs prodding my ribs. The path snakes over a series of cold-looking mountain streams before, ultimately, a small village of A-frame huts breaks through the fog. Horombo is a relatively well-maintained staging point. Most hikers attempting the Coca-Cola ascent arrive here during their second or third day on the trail. A sign announcing an elevation of 12,100 feet greets me in front of a thin radio transmitter that provides Horombo's main form of communication to the outside world.

Unsure of what to do next, I walk into one of the huts and find myself in a dining room where a Swedish expedition is enjoying their dinner. Every head turns to the half-naked man in their midst. There are smiles and someone in back shouts, "You must think you're Wim Hof," in what I take to be jest.

"He's around here somewhere. Have you seen him?" I ask, genuinely curious as to where the leader of our expedition has disappeared to. The man who yelled jumps up and almost runs to me. He is holding a camera in his hand. He starts to whisper in excitement. "Really? You're here with Wim Hof? He's an idol of mine. I can't believe you walked up here like this." I must look like a creature from another planet. But I have to admit that it feels good to be admired just for taking off my shirt. The

Swede snaps a few photos of me outside and asks if it is possible for me to introduce him to Hof.

The rest of the group arrives over the next 2 hours, and each shirtless mountaineer receives a hero's welcome. We are clearly the talk of the camp, and we all toast with hot tea and a hot meal as the rest of the group straggles in one by one. It isn't until Sylvia—the proprietor of the Dutch marijuana shops—arrives well past dark that I get a taste of what lies ahead. The climb has worn her out, and she's in tears when she crosses the cabin's threshold. She says that her head is pounding, and she falls against the wall with her head in her hands.

A crowd gathers. Hof appears from somewhere and folds her into a paternal hug, warming her up at the same time as he whispers something in her ear. Someone measures her oxygen and it is critically low, running at less than 50 percent saturation. Hof bends over so he can look her in the eyes and they start to breathe together. She focuses on him and after what must be 300 breaths regains some control. Her tears dry up but she's worn out. Meanwhile a couple of guides confer with each other about what they should do next. If someone dies it will be their necks on the line. No one wants to hire a guide who let someone die on their watch. In about 30 minutes she's able to get her oxygen saturation back above 90 percent, but there's no telling what might happen if she climbs farther. Someone decides that she'll have to go back. She's the first casualty of the hike.

At night the group sleeps head to toe in a long series of bunk beds that line the length of the longhouse. The windows are open to the night air in order to ventilate the CO_2 that builds up in a small space with 30 people. At 3:00 a.m. an alarm goes off reminding us all to check our oxygen levels and bring them back up to 90 percent. It's hard to get back to sleep, but at 4:30 Hof is full of energy and almost shouting that we need to get moving or miss our window to the summit. The hike will take 11 hours in progressively thinner air, and no one besides Hof feels motivated to get going this long before the sunrise.

Breakfast is a bland portion of porridge and eggs. It's far from deli-cious, but eating is important so I try to pack in as much as possible in 5 minutes. That's when I notice van Winkle—the tall Dutchman who was having trouble earlier—on the porch doubled over the banister. He's heaving hard and spewing what seems to be gallons of a yellowish liquid over the edge. When he hears about van Winkle's condition, Hof slaps him on the back and urges him onward. "This trip will separate out the men from the boys," he says in a comment that might have been intended to be motivational but instead falls a little flat. For all of his enthusiasm and inspirational achievements, Hof sometimes lacks conventional social skills.

Van Winkle looks at Hof and shakes his head. "My choice seems to be to go down or die," he groans back. His climbing partner nods in agreement. Hof seems disappointed.

"Fine," he says. Then he calls to the rest of the group to get ready. It's time to go up.

Hof is an almost impossible figure to dissect. On one hand he has something special—physical and mental abilities that seem to open the door to untold reservoirs of strength. He is a prophet whose message is spreading across the world in viral videos and scientific journals. He is a man whose love for all things human seems to know no bounds. He is also a madman who can be so solipsistically focused on his own abilities that he lacks the empathy to see limits in other people. When he aban-dons the group in search of a record, or when he delves into a 45-minute speech on how the cold and breathing can bring him deep into his own physiology, Hof can fall into such a fit of self-centeredness that there isn't space for anything else. At sea level, or in a temperate zone, it's a person-ality foible that is easy enough to overlook. But those qualities can turn dangerous on a mountain where the lives of almost 30 people hang in the balance. So when the group leaves that morning in a line that is two people shorter than it was the previous morning, I wonder: Am I follow-ing Hof the prophet, or Hof the madman?

As we trudge out in the darkness with only a line of headlamps illuminating the rocky dirt path, my choice of clothing telegraphs my unconscious answer to the question. Of the 27 of us only one isn't wearing a shirt. It's cold, but I'm shirtless because I'm not yet willing to abandon my faith in the method. I think warm thoughts—of a fire burning in my belly—as I walk forward willing heat into my naked torso. Even Hof is wearing a T-shirt and a blanket draped over his shoulders.

Our aluminum poles and hiking boots crunch the dirt as we move ever northward up a volcanic chunk of rock that claims the lives of at least eight mountaineers a year.[2] Our breathing is harsh and rhythmic, as if we are trapped in a room from which the air is being sucked away. It is as if any lungful might be our last. But we trudge forward through the darkness in focused unison until fingers of orange light grab at the horizon to pull away the night. It is only then that the dark outline of a mountain peak begins to define itself. At first it is only a dark, purple absence of stars in a pinprick sky. As the heavens shrug off night's embrace the sunlight sets the glacier ablaze like a beacon.

The summit of Kilimanjaro.

The tallest mountain in Africa rises up out of the sun-drenched savanna to a place high above the clouds. There, winds topping a hundred miles an hour scour a glacier that is probably the only indigenous ice on the continent. For the past 20 hours the peak was hidden behind clouds and the mountain's own towering foothills. Now in view, the massive slab of igneous rock is no longer an idea we might conjure up in our minds, but a real-life obstacle. The gradual, 15.5-mile rise from the park gate comes to an abrupt halt at the base of the cone that rockets upward into a barren and inhospitable wasteland. Devoid of life and home to only a moon-like base camp, we hope to summit on almost no food, little sleep, and most strikingly for me, no cold weather gear.

One of the guides that we'd hired in Moshi watches me warily from

---

2   Accurate casualty counts on Kilimanjaro are surprisingly difficult to obtain. Various sources indicate the number is between three and 10 reported deaths a year, while the actual number may be as many as 30, since some cases may not make official records. Either way, there were 1,000 annual evacuations from the mountain top.

beneath his full thermal getup. "Please put something on," he says, confused by the rash display of skin. It's a sensible request. Even with the sun tempting the sky, the air temperature is well below freezing and it is only going to get colder as we make our way higher. But the request is also an affront to why I'm on the mountain in the first place.

So I push his words out of my mind and try not to think of how the others in the group are dressed. I suck in a cool breath of air and focus my eyes on the blazing orange rock in front of me. I exhale something that sounds like a low guttural roar, like a dragon just waking from a hundred year slumber. I feel the energy begin to build. The rhythm of the air quickens. My toes start to tingle inside my hiking boots. The world starts to brighten in my vision as if there are two dawns working at the same time—one tied to the rising of the sun, the other in the depths of my own mind. A coil of heat starts behind my ears like a lit fuse. It arcs across my shoulders and down the curve of my spine. There's no point in checking the temperature. It's cold and I'm only just beginning to sweat.

Or so I think. In another mile we reach the saddle—a low, sloping curve of land that links a small castle-like volcanic peak just across from Kilimanjaro's main cone. The shape of the land acts as a sort of wind tunnel that accelerates the gusts into gales. The taut straps on my backpack sing like a stringed instrument. We lope past a small circle of rocks that demarcate the border of a helipad that doctors use to evacuate victims of altitude sickness. Someone calls for the group to stop so that they can relieve themselves in a small wooden outhouse. The temperature hovers around 10 degrees, but the gusts create a wind chill that brings the temperature down to the negative 30s, a temperature that the tables I was given by Castellani say could frostbite exposed flesh in as little as 15 minutes.

The Wim Hof Method does not make a person impervious to the elements. Pausing the march for even a few moments in such an exposed area makes it all the more difficult for my body to generate the amount of heat I need to fight the cold. As we mill around the helipad I feel my body temperature sink. I know that it will take much more energy to get warm again. The break is only 5 minutes or so but when we start up again I can feel that I'm in trouble. Hof pushes past the front of the group and presses on. I wonder if his impetuousness is a symptom of needing to maintain a constant pace, that his own body temperature isn't maintained by his mental control alone but also by the slow, steady work of his muscles.

I watch the blanket on his back push farther into the distance as I dip toward the nadir of the mountain's saddle. The wind howls past me at some ungodly pace, and I start to think that it's more reasonable to put on a few layers here than to risk not reaching the summit at all. So I drop my backpack and pull out a thin merino wool shirt and wool sweater that will provide my skin with a little protection. They're not much of a barrier against the harshest gusts, but they are all I need to feel safe. With a burst of renewed energy I pick up the pace until I catch Hof. The rest of the group further segments behind us.

In some sort of optical illusion, the vastness of the landscape makes distances look much shorter than they really are. Though the way forward is mostly flat, progress is insufferable. A trick in optics takes what looks like a 15-minute journey and stretches it into a grueling hour. It's been 5 hours since we left Horombo, but we're so spread out now that the last people in the group won't reach our next waypoint for another 45 minutes. The mirage psychically drains the group of its motivation.

The waypoint ahead of us is a place called Kibo, the base camp for everyone who gets to the summit on the Coca-Cola route. It represents our last prospect for rest before attempting the summit. From there, Kilimanjaro makes a sharp upward incline. Kibo itself is far above the point where any plants are able to grow. It is a barren outpost that features a handful of stone buildings, a kitchen, and some quarters where rangers can stay for a few weeks at a time to monitor the constant flow of mountaineers who come through. We find a stone room with a collection of basic wooden benches and tables, and try to warm up before our final ascent. Ordinarily hikers will stop here overnight to acclimatize and rest, but our plan is to climb 6 more hours to the peak. It is not quite 11 a.m., and the clock is still on our side. Dennis Bernaerts, the Cross-Fitter and Ironman who hangs on Hof's words like gospel, is shivering in the corner. When I donned a sweater in the saddle he was inspired to take off his shell. He was half naked during the harshest part of the hike. Now he's breathing hard trying to warm up. He's pale and his lips are blue. I pull a green puffy jacket from my bag and give it to him to wear.

Hof lies back on one of the tables and closes his eyes. They roll back in his sockets and flutter. He turns red and when he opens his eyes he is once again bursting with energy. One of the problems with following someone with superpowers is that my body doesn't obey the same rules of rest and recovery that his does. And now that we're close to the summit his mind is focused only on setting a record.

Everyone else is recovering and waiting for a hot lunch that we'd been told was on the way. Hof checks the time. It's 11:40. Then he shouts, "Okay, everyone. We go in twenty minutes. We're not on the mountain because we want to eat. We're here to make the summit." The message lands flat in a room full of confused faces. I'm not the only one to mutter an obscenity. Sensing insubordination, Hof revises his message 3 minutes later. "Never mind rest," he says. "We go now and make a new record!" He launches himself out of the room and walks down the wide path between Kibo's humble buildings and into a collection of mountain guides. Mike Nelson, the guide who helped retrieve Mallory's effects off Everest, tries to block Hof's path.

"You can't go now," he tries to command Hof. But the gnome-like leader of our expedition simply brushes past him. He heads up the mountain towing two Belgian ladies who had come on the trip to enhance their corporate leadership skills. I watch the movement of his neon orange shoes from the doorway while trying to decide whether I have a duty to follow the leader or if I should participate in what seems like a well-justified mutiny. I watch him grow smaller on the mountain. About a third of a mile from the camp he turns and starts yelling something. I can't decipher his words. It could be a last ditch effort to restore morale. It might just be a string of insults.

I'm pissed. This is not the trip that I signed up for, and my faith in Hof is quickly dwindling to a low point. But part of me also wants to know what happens next. So I come up with a compromise. I rifle through my bag and remove every warm-weather article of clothing that I'd packed along. I throw on wind pants and a thermal top. I cover myself in a sweater and retrieve my green puffy jacket that I was only going to use in an emergency. I finish it off with my yellow rain coat. I'm positively toasty in the layers. Not because I need all this clothing, but to protest against Hof's impetuousness. Fuck the method. Fuck Hof. Thus armored I walk out of Kibo camp and yell up to Hof, who is now in the middle of some sort of argument with the Belgians.

"Hold up!" I yell. My shout echoes off the canyon's rock walls. I walk past the guides, who are arguing with each other in Swahili. If something goes wrong with the ascent, they know that they will share in the blame and consequences. Only Bernaerts follows in my footsteps.

The Belgian ladies have turned their backs on Hof. He's all alone, standing in his shorts and blanket a few hundred meters above us. "Why not give us another hour?" I yell up the mountain. "Even the guides won't follow you." Perhaps the wind eats my words and sends them down the mountain, but Hof hears them as an insult.

"Don't you dare challenge me, Scott!" he roars back with the wind howling an angry chorus.

"How do I know that if I follow you that you won't leave me behind on the mountain like you have everyone else on this trip?" These words seem to reach him. Alone on a ledge above me there's a realization going on. Yes, he can make the summit, but is it worth it if no one follows? It takes a while. Maybe 15 seconds. Maybe a minute. But he calls back, his voice softer.

"I won't leave you." He waits another beat. "I promise."

It's not much, but in the moment it's enough. In a few minutes Bernaerts and I fall in behind him. We gaze back at Kibo. It looks even more desolate from above. We can see the guides bickering and the Belgian ladies gesturing in our direction, but it's impossible to know what they are saying.

"Maybe they will come," Hof says with bitter optimism. But the three of us know that it isn't very likely.

So he turns his back and takes a small step up the mountain. And then another. We follow behind. *Pole, pole.*

I watch his shoes and blue, bird-emblazoned swim-trunks for the next half hour, mostly in silence. There's just the sound of my aluminum poles followed by the crunch of boots and constant breathing. The rhythm is almost hypnotic except for a question that I just can't get out of my mind: Why would I follow a madman up a mountain? Hof is many things, but a leader of men is certainly not one of them. I have my doubts about whether he will keep to his word, or if once we overcome one rise or another he will push on while my own strength flags. And it isn't until I see his shoe slip slightly on the gravel, a minor misstep in which even his own profound fortitude falters for an instant, that I realize that I'm not following a madman. I'm not following a prophet. I'm not even following Wim Hof. I've come up the mountain after years of working on my own biology, one step after another. I'm following my own rules, my own relation to the group, and even the mountain itself. I don't care if we make or break a record. I want to know my own limits.

The thought adds a certain jaunt to the next several steps. I'm not angry at him anymore. Suddenly my layers of clothing and the protest they represent feel misplaced. I drop my bag to the ground and strip down to my waist. The wind isn't as fierce on this slope of the mountain and even going slowly I'm working hard enough to keep warm. Stripping down feels like a small triumph. If I do layer up again it won't be because of anything Hof says or does. Insane or not, this is my challenge. I'll earn the summit or fall on my own power.

Hof keeps his hands clasped behind his back, focusing on every footstep on the path upward. The lip of the volcanic cone is still more than 2,500 vertical feet above us, but every step gets us that much closer. The slope dips slightly, eclipsing Kibo from view, but I look back anyway and see a smeared red silhouette barreling up the mountain behind us. He's gaining fast, taking two steps for every one of ours. I tell the others. We squint into the distance and recognize him as Salim Hamis Ngonye, one of youngest guides in the expedition. Whatever the dissent was below, he was either sent up to keep an eye on us or had broken away on his own. Even so, it still takes him 45 minutes to catch up after we stop to take a break on a pile of rocks.

We don't have much food with us, having skipped our resupply at Kibo, but I fish out two granola bars from my bag. Bernaerts uncovers some sort of medicinal-tasting performance gel and we share the small meal. We're so high up that it proves somewhat difficult to chew and breathe at the same time. Each mastication competes for the space that we could use to draw in air. Our oxygen readings are all above 90 except for Hamis Ngonye's, who comes in below 70 percent. He hasn't been using the breathing method, and I wonder if simply being on the mountain every month has habituated him and the other guides to functioning well even while at a deficit.

Just like with the various ways that a body might choose to heat itself, there are a wide array of strategies and techniques that our bodies use to manage the lack of oxygen at high altitude, but for me the breathing method is effective. We continue upward in a line of four people with Hamis Ngonye in the front. Sometimes the rhythm of the march lulls me out of conscious breathing. My mind starts to wander and I only take in as much air as my mind might see fit and forget to lead my breath. That's when the creep of altitude sets in. The world dims imperceptibly and every footstep seems to fall just a little heavier. Then, as I realize I'm fading, I start 30 rapid breaths and the world brightens up as easily as if I were taking off sunglasses. My steps become lighter and I have the energy to continue. At this height the mountain makes me keenly aware of my body. I even feel as if I can sense the energy from the granola bar coursing through me as the mitochondria in my cells greedily consume the sugars in exchange for brute muscle power.

We pass a small cave and then a landing before we make a long zig-zagging traverse over what feels like volcanic sand. Craning my neck, I make out a white patch of snow above that signals the rim of the crater. Gilman's Point isn't the true summit of Kilimanjaro, but it is the point where you can cross over the top of the mountain and start going down the other side. The true peak—Uhuru—is another hour and a half around the rim of the crater along with another few hundred feet of elevation. Hof, Bernaerts, and I agree that for our purposes, Gilman's is enough. We don't need to go any farther to prove anything to ourselves. An additional 3 hours of walking there and back just doesn't seem worth it for such a modest gain. Even so, the distance to Gilman's still induces vertigo, as my vision telescopes in and out on the snow and a dark cloud that has started to engulf the summit. I opt to keep my head down.

In another half hour we reach a stretch where the gravel path gives way to boulders—a stretch called Jamaica Rocks—forcing us to go hand over hand in places. It's the most technical part of the climb, and it's made worse by the sky suddenly opening up and sending down a flurry of snowflakes and wind. It's about 5 degrees Fahrenheit, but later I would calculate that the magnifying effect of the wind on skin brings the real temperature down to minus 31. That's enough to cause frostbite in just a few exposed minutes, though I've been shirtless for the last several hours. I watch as Hof stumbles on a set of rocks, sending his hand down onto the path. It is apparent that we're all reaching our physical limits. Between the speed of the ascent, the cold, and the wind chill I know I need to focus my energies on just putting one foot in front of another. I put my sweater back on and awkwardly plant my poles onto the rock face as I prepare for the last push.

There are many great mountaineers in the world, but I am not one of them. Hof and Hamis Ngonye are both experienced climbers, but even the moderate boulder scrambling is a test of my abilities. The poles get in the way when I need to crawl, and it's hard for me to balance without them when I'm standing. As the three other climbers seem to vault airily over one 6-foot ledge, I grasp the porous rock with my fingers and desperately try to find a hold somewhere below me with my toe. I slip and my body crashes forward, and I splay out with my arms akimbo. Hof watches my cacophony of limbs and says to Bernaerts or Hamis Ngonye, or maybe just to the air at large, "He's fighting himself now."

I'm too drained to take the comment as anything but an insult, but it's also true. I am faltering on the mountain. I am fighting my own muscles and mind as much as I am the boulder that my body is splayed across. But goddammit, I'm going to make it to the lip of this mountain if it kills me. And this time, when I look up, I realize that our goal is just a couple hundred feet ahead. I can fight myself or I can fight the mountain. At this point it isn't even a choice anymore.

Then, before I know it, we're there. Breathless at the top of the African continent. Maybe just a little shy of the true peak, but when we check our watches and subtract away our departure time we calculate that we've more than beaten our 30-hour goal; we've crushed it. It has been only 28 hours and 6 minutes since we left the park entrance. To put that in perspective, when Hof's group did the same trek the year before, it took them 42 hours to reach Gilman's Point. To the best of our knowledge it is the fastest-ever ascent to Gilman's Point by a group of people. Somewhere below us the rest of the group is on their way up. They will reach the same point in about 2 hours. Altogether, 22 out of 29 will make it to the top without succumbing to AMS. Though Castellani and the US Army predicted mostly failure, we will achieve a 75 percent expedition success rate.

And so I breathe in that success in Kilimanjaro's thinnest air. The wind whips past the large, wooden sign that marks the elevation, and I start to feel warm again. I take another 30 breaths and I'm hot. So I take off my shirt again to feel the weather on my skin. Even if the wind only speaks in the language of frostbite, it feels like victory. I'm not on the mountain anymore. In this moment, I am the mountain.

EPILOGUE

COLD COMFORT

IT'S BEEN 6 months since I last met with Rob Pickels at the Boulder Center for Sports Medicine, and things couldn't feel any more different. For one, the center has cut a deal with the University of Colorado Boulder and moved into a brand new facility attached to the school's sports arena. The state-of-the-art office overflows with meaty-looking football players and spindly endurance athletes. The upgrade comes with a new name, the CU Sports Medicine and Performance Center. But other than those cosmetic changes the test that Pickels will do on me is essentially identical to the one I did with him the previous summer. There's no doubt that I feel different, though.

After a few years of exploring ways to use the environment to peek into the inner workings of the body, I've learned a thing or two about my own limits in the world. I know that every person—even people who have jellyfish spirit animals—hide vast wells of inner strength. The secret to cracking into our inner biology is as easy as leaving our comfort zones and seeking out just enough environmental stress to make us stronger. Exposure to cold helps reconfigure the cardiovascular system and combat autoimmune malfunctions. It is also a pretty darned good method for simply losing weight. These are all things that I've seen in myself and in hundreds of other people who have had the will to delve deeply into their own evolutionary physiologies. More profound than that, however, is the intrinsic understanding that humans are not just bodies bounded by the barrier of their skin; we are part of the environment that we inhabit.

Just a week off Kilimanjaro, I figure that I'm in the best shape of my life. My once-expanding waistline has tucked back in. An old pair of jeans featured a 36-inch waistband, but my most recent purchase was a size 31. I used to suffer from occasional canker sores—painful dime-sized autoimmune lesions that I'd been getting since I was 2 years old—but I haven't had one since I started taking cold showers.

For the last year I'd been taking on a few big challenges, but my weekly exercise routine is pretty much unchanged. I run outside shirt-less, no matter the weather, two or three times a week. In the summer I go through countless tubes of sunscreen, and in the winter I sweat on 3-mile runs around a lake by my house. After working with Brian Mac-Kenzie at his facility in Southern California, I start throwing in a few HIIT sprints for good measure. All told, including morning breathing routines, my fitness regimen takes about 3 hours out of my week.

Before he starts hooking me up to breathing tubes and sharpening his medical lancets, I tell Pickels about my routines. This man who spends most of his day with top-tier athletes seems to chuckle inside. He isn't expecting much. So when he starts me on the treadmill once again and incrementally increases the inclination and speed of every stage it is more to humor me than it is to declare a new fitness paradigm. The first few minutes aren't much of a challenge, but as he pushes the machine faster and higher I begin to sweat and the rubber facemask starts to pull me awkwardly to the side.

"You're doing great. You got this," he says as he pokes my finger with a small lancet in order to draw some blood. He feeds the sample into the machine and then increases the speed and inclination until it feels like I am running up Kilimanjaro again.

"You've done another stage more than last summer. Keep it up!" he says. I smile beneath the breathing apparatus and give him a thumbs up. I run on until my lungs burn and my mind screams to stop. Then I put my feet on the side of the machine and let the belt spin beneath me. In that moment of acquiescence, I think that I might have had a little more energy left in reserve. Maybe there is one more stage I could have done. I tell Pickels as much. "Everyone says that," he replies.

There are a lot of numbers to crunch, and Pickels tells me to come back a few days later for the results.

A week turns into a month, and in that time I continue my daily routines. Eventually we make an appointment that sticks and early one morning in February I drive out to meet him. He greets me in the lobby of the center with a vigorous handshake, the first sign that the results are more interesting than he had expected.

We sit down in a small conference room with a TV on the wall where he can display the 40 pages of data points and complex math. He talks fast as he delves into describing the bicarbonate cycle, metabolic actions, and then he outlines a few equations that make my head start spinning. It is more than I can understand, so I ask him to bring it all down into regular English.

"Big picture: You're a lot more efficient with how you use energy than you were when I tested you in the summer. It's like you've added seven hours of exercise to your routine every week," he says. In addition to running longer on the treadmill, I burned more fat and fewer carbohydrates—meaning that I'd trained my body to use less sugars and more stored energy.

He shows me a few graphs to illustrate the points. Two charts testing fat and CHO oxidation rates over time show two intersecting lines: a dark line for carbohydrates and a lighter line for fat. Back in May the two lines had intersected before the third stage on the treadmill. When I'd started the run in May I burned fat almost instantly, but the levels plummeted as the workload increased. My body had to burn more carbohydrates to compensate until it finally flamed out after six stages. In the second chart, from January, the two lines intersected in the fifth stage and I ran for seven full stages. This meant that I burned fat for the bulk of the workout.

"Which is huge," he says.

Other things didn't change, though. In both studies my lactate production was three times higher than the average adult's. Lactate is a byproduct of physical exertion and often correlates with long recovery times after intense workouts. It's the acidic byproduct that makes you sore after running a race. Given the change in fat burning, Pickels expected something more here, but he shrugs.

"Everyone's physiology is different," he says.

Overall, the changes in my physiology are too big to be accounted for by simply one climb up a big mountain—even something as challenging as Mount Kilimanjaro—but instead mark a prolonged change in the way my body uses energy. Maybe, he posits, it has to do with how I'd simply made myself comfortable in the cold. "Being cold forces you to use more energy. It's like a low-grade passive workout for your whole cardiovascular system," he muses. Research, he notes, indicates that cold exposure increases mitochondrial formation throughout the body, which would mean more overall aerobic power.

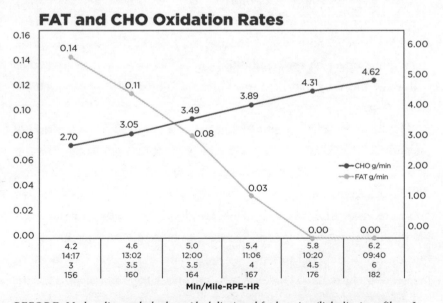

FAT and CHO Oxidation Rates

BEFORE: My baseline carbohydrate (dark line) and fat burning (light line) profile as I reached my VO₂ max, as recorded on May 1, 2015, before starting the Wim Hof Method on a regular basis. I managed six stages, and the two lines intersected between the second and third stage (meaning I started burning more carbs and fewer fats at that point).

The positive results are all the more impressive because my actual workout routine hasn't changed more drastically than adding cold showers and the morning breathing sessions. When I first met Hof in Poland almost 4 years earlier I weighed 210 pounds. Now I'm 178. More than that, I'd proven at least a little bit that I could handle altitude and freezing cold. I'd also climbed the tallest mountain in Africa.

On one level I am incredibly pleased with how the tests came out. But I think the reasons behind the improved results run deeper than cold exposure and breathing exercises alone. They stem from a deep connection that I am forming with the world around me. It's something that I hope anyone can at least try for themselves.

Every nerve ending that connects our skin to our brain yearns to understand the world around it in order to help us make appropriate decisions about how to best go on living. Most of these signals happen

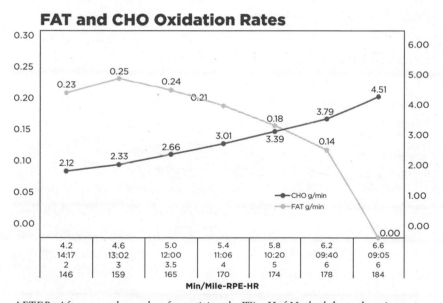

AFTER: After several months of practicing the Wim Hof Method, but otherwise keeping my exercise routines the same, I managed to radically change my metabolism. This graph of my results from the January 27, 2016, test indicates I managed seven stages of increased exertion and, more importantly, continued to burn primarily fat for five stages of the test. To quote Rob Pickels: "Big picture: You're a lot more efficient with how you use energy than you were when I tested you in the summer. It's like you'd added seven hours of exercise to your routine every week."

below our conscious understanding in the relays of our most lizard-like brain structures. These pathways are the programming that has been passed down by countless generations of ancestors all the way back to the very inception of mammalian life itself. The biological relationship between life and its environment is the most ancient transfer of knowledge that any human can ever experience. It's so ingrained in our very beings that the language of evolutionary programming isn't a thought, it's a feeling: a shiver, a rush of blood, or a quickening of the senses. Just a hundred years ago most of us knew the language of shivering. A thousand years before that our bodies tracked with the seasons. Ten thousand or more years before that our species migrated between continents on rafts of seaweed and surmounted mountains in little more than animal skins and leather soles. Those ancestors probably didn't think of themselves as different from the environment at all. They knew what we are learning again today. That we are all just here. Surviving. Together.

Even though our nervous systems crave connection to the world they evolved in, the tendency in the modern era is to think of humanity as fundamentally different than everything else. We insist on being bounded by our bodies, our property, our kinship network, and our social media profiles; and so we hold the rest of the planet at arm's length. The combined effect of billions of people sharing that same atomistic mind-set has set the world off-kilter. As I was writing this book, 2015 and 2016 were among the hottest years ever recorded. In the years to come, actually finding cold will become more difficult as winters become milder and summers all the more scorching. There may come a time when finding an ice-covered lake in Boston will be impossible. On a global scale, climate change is another signifier that humans are inextricably part of the environment. Humanity has used its cerebral ingenuity to extract the stored energy of the planet for its own purposes. The collective action of our carbon-dioxide emissions, rampant pollution from the industrial revolution onward, and the mass extinction of animal life are just a few iconic ways that we've steered the environment toward our own, ultimately destructive, ends. The process has been mostly unconscious—no one planned the destruction of the planet—but the results are so massive that they almost seem calculated. In a way, you could say that humanity itself is now the world's conscious nervous system. And, just as our own bodies have grown weaker as a result of clinging to comfort and using our gadgets to appease our inner drive for homeostasis, the alterations we've made to the planet through technology have undermined its fundamental equilibrium.

Of course modern technology has done good as well as harm for our species. Human life expectancy is far higher now than at any time in history. While our Paleolithic ancestors might have had better teeth, stronger muscles, and hardier immune systems, they also faced lethal challenges that cut most of them down by their mid-thirties. Infant and maternal mortality was obscenely high. Epidemics repeatedly wiped out entire cities during the Middle Ages. All told, I'd rather be alive now than at any other time in history. Technology gives us superpowers, and so what if I can't always find my way around a city without my phone's GPS? I've had the amazing opportunity to travel to far-flung parts of the planet in a way that would have been unthinkable even a generation ago—flitting from Colorado to Moshi, to Amsterdam and London, all in just a matter of months. I've had my own limits tested in state-of-the-art laboratories and with some of the most inspiring athletes on the planet. I'm lucky. Then again, we all are.

I'm not going to make an appeal for anyone to try to get up and save the planet or alter the course of human history. I will, however, note that every one of us has an opportunity right now to connect to the world around us. If you've been wrapped in a thermogenic cocoon for your whole life, then your nervous system is aching for input. All you need to do is get a little bit outside of your comfort zone and try something out of the ordinary. Try finding comfort in the cold. You have nothing to lose. Just breathe.

NOTE ON THE CHRONOLOGY

WHERE POSSIBLE I have attempted to present this book in chrono-logical order. However there are several areas where I have taken some liberty with the timeline in order to preserve narrative consistency. In a few places I have condensed quotes and images into single scenes, which, in fact, occurred over the course of several days. Perhaps more signifi-cantly, astute readers may also notice that my ascent of Kilimanjaro at the beginning of January 2016 occurred several weeks before I ran Tough Guy in the UK, even though the chapters are not presented in that order.

ACKNOWLEDGMENTS

NOTHING IN THIS book would have been possible if I hadn't tried to debunk the prophet and madman Wim Hof. His pioneering impulses allowed me to think about my own body in an entirely new way, and it is hard to express how much gratitude and respect I have for him. I'm glad that I was wrong about his method. It's also been a constant inspiration to witness the work of other pioneers who have used the environment to manipulate their own hidden biology, from Laird Hamilton and his wife, Gabrielle Reece, to Brian MacKenzie.

Of course I'm also indebted to the diligent work of my editorial team. First to my editor at Rodale, Mark Weinstein, who chaperoned me through every stage of this book's development and advocated for the project within the walls of an unfamiliar publishing house. Also my literary agent Laura Nolan, who helped focus the initial idea from a magazine story in *Playboy* into a coherent proposal. I am also fortunate that Mary Ann Naples, who was my literary agent on my first book, *The Red Market,* had become the publisher at Rodale when it came time to pitch a book on an esoteric Dutch guru who had a plan to rethink humanity's relationship with the cold. Naples has since moved on to be a vice president at Disney Publishing, but Rodale's new publisher, Gail Gonzales, has become a strong advocate for this book. Along the way I also had constant support from the Schuster Institute for Investigative Journalism, and I would especially like to thank Kiara Tringali, who helped transcribe endless hours of interviews, as well as Florence Graves and Lisa Button, who helped advise me on the project from its earliest stages.

While I had the opportunity to synthesize a great deal of scientific research, I am constantly humbled by the diligent efforts of researchers who have spent decades documenting our hidden human biology. Among the scientists who graciously lent me their knowledge were Ray Cronise, Aaron Cypess, Peter Pickkers, Matthijs Kox, John Castellani, Marc Kissel, Maria Kozhenikov, Daniel Lieberman, Tony Gustafsson, Shaun Morrison, Kevin Phillips, and Richard Wrangham.

I had the great pleasure of spending a few weeks in Holland, where I got to meet many people who have been helped by the Wim Hof Method including Enahm Hof, Geert Buijze, Henk van den Bergh, Manely Ellamo, Henk Emmink, Kasper van der Meulen, Bart Pronk, and Isabelle Hof. And, 2 years before that in Poland, I met Hans Spaans, Vladamir Stojakovic, Janis Kuze, Ashley Johnson, and Andrew Lescelius.

In Los Angeles I was honored to brush shoulders with actors Orlando Bloom and John C. McGinley as they scooted along the surface of a 12-foot-deep pool while carrying ungodly amounts of weight in their hands. I was also fortunate to meet Darien Olien and study his investigations into human nutrition. This book would not be nearly as beautiful without the photographs that Chris DeLorenzo took while both in the pool and on solid ground. I was pleased to also be able to spend a few days with yoga savant Sati Ah in Long Beach and learn a bit of surfer gossip from Sarah Spivack LaRosa and her husband David in Venice.

In the United Kingdom, Ed and Will Gamester showed me how dressing in masks and getting far too drunk is not necessarily the worst way to get prepared for an obstacle course race, while Mr. Mouse shared with me excellent hospitality at his Home for Unfortunates. Photos of the event were expertly documented by previous Tough Guy winner James Appleton, images that would never have been taken if he hadn't broken his hand a few weeks earlier. Also along on that trip was Niklas Halle'n, who captured a beautiful image of Ed performing a high-flying kick on Madhi Malik inside the ring of a British burlesque club.

As with many journalistic endeavors, this work also builds on the work of other writers. Scott Keneally introduced me to the world of obstacle course racing (and I highly recommend his film *Rise of the Suf-*

ferfests), while Steven Leckart at *Wired* helped me frame some of my ideas about BAT and cold training. Jayme Moye in Boulder introduced me to the CU Sports Medicine and Performance Center so that I could track my own fitness results.

Getting to the top of Kilimanjaro would have been that much more difficult if Salim Hamis Ngonye hadn't followed three impetuous adventurers up the mountain, while the veteran guide, Mike Nelson, was also probably right to urge caution.

In Boston I had a chance to reconnect with old friends Siiri Morley, Jeremy Ogusky, and Claire Beckett, who generously lent me a place to crash while I investigated the city's brew of cold and wet winter. They got me ready to meet Bojan Mandaric and the good people at November Project, who urged me up countless flights of stairs at Harvard Stadium.

I was also energized to surf the cold waters of Cape Cod with legendary board shaper Shawn Vecchione, who so impressed me with the effectiveness of modern-day wetsuits that I decided to cut a chapter on winter surfing.

I also enjoyed early morning workouts with the November Project tribe in Denver and, of course, Jeff Vahrenwald's encouragement and sense of humor throughout the writing process. And thanks to Santosh MP who cut an amazing book trailer.

I have been fortunate to have a loving and supportive family throughout my career—some of whom have even jumped into freezing water with me.

Above all else, however, this book would have been a hollow affair without the constant support and advice of my wife, Laura Krantz, who not only accompanied me on a few of these adventures but who gets up many mornings with me to breathe and take cold showers as we explore this method together. She makes every day a little better than the one before it.

WORKS REFERENCED

Benson, Herbert, et al. "Body temperature changes during the practice of g Tum-mo yoga." *Nature* 295 (January 21, 1982).

Berger, Robert. "Nazi science—The Dachau hypothermia experiments." *New England Journal of Medicine* 322 (May 17, 1990).

Buijze, Geert, and Maria Hopman. "Controlled hyperventilation after training may accelerate altitude acclimatization." *Wilderness & Environmental Medicine* 25 (2014): 484–494.

Cannon, Barbara, and Jan Nedergaard. "Nonshivering thermogenesis and its adequate measurement in metabolic studies." *Journal of Experimental Biology* 214 (2011): 242–253.

Chvetzoff, Gisele, and Ian Tannock. "Placebo effects in oncology." *Journal of the National Cancer Institute* 95, no. 1 (2003): 19–29.

Cronise, Ray, et al. "The "metabolic winter" hypothesis: A cause of the current epidemics of obesity and cardiometabolic disease." *Metabolic Syndrome and Related Disorders* 12, no. 7 (2014).

Cypess, Aaron, et al. "Identification and importance of brown adipose tissue in adult humans." *New England Journal of Medicine* 360 (April 9, 2009): 1509–1517.

Darwin, Charles. *The Voyage of the Beagle.* Project Gutenberg (1839).

De Lorenzo, F., et al. "Cold adaptation and the seasonal distribution of acute myocardial infraction." *QJM* 92 (1999): 747–751.

Devlin, Maureen. "The 'skinny' on brown fat, obesity, and bone." *Yearbook of Physical Anthropology* 156 (2015): 98–115.

Gillen, Jenna, et al. "Twelve weeks of sprint interval training improves indices of cardiometabolic health similar to traditional endurance training despite a five-fold lower exercise volume and time commitment." *PLOS ONE* (April 26, 2016).

Hanssen, M. J., et al. "Short-term cold acclimation improves insulin sensitivity in patients with type 2 diabetes mellitus." *Nature Medicine* (July 6, 2015).

Hof, Isabelle. *The Wim Hof Method Explained*. Innerfire (2011).

Hof, Wim, and Justin Rosales. *Becoming the Iceman: Pushing Past Perceived Limits*. Mill City Press (2012).

Kaciuba-Uscilko, Hanna, and John Greenleaf. "Acclimatization to cold in humans." National Aeronautics and Space Administration, April 1989.

Kipnis, Jonathan, et al. "T cell deficiency leads to cognitive dysfunction: Implications for the therapeutic vaccination for schizophrenia and other psychiatric conditions." *Proceedings of the National Academy of Sciences* 102, no. 39 (September 2005).

Kox, Matthijs, et al. "The influence of concentration/meditation on autonomic nervous system activity and the innate immune response: A case study." *Psychosomatic Medicine* 74 (2012): 489–494.

Kox, Matthijs, et al. "Voluntary activation of the sympathetic nervous system and attenuation of the innate immune response in humans." *Proceedings of the National Academy of Sciences* (May 2014).

Kozak, Leslie. "Brown fat and the myth of diet-induced thermogenesis." *Cell Metabolism* (April 7, 2011).

Kozhevnikov, Maria. "Neurocognitive and somatic components of temperature increases during g-Tummo meditation: Legend and reality." *PLOS ONE* 8, no. 3 (March 29, 2013).

Leckert, Steven. "Hot trend: Tapping the power of cold to lose weight." *Wired* (February 13, 2013).

Leonard, William. "Physiological adaptations to environmental stressors." *Basics in Human Evolution*. Compiled by Michael P. Muehlenbein Elsevier, 2015.

Leonard, William, et al. "Metabolic adaptation in indigenous Siberian populations." *Annual Review of Anthropology* 34 (2005): 457–471.

Lieberman, Daniel. *The Story of the Human Body: Evolution, Health, and Disease*. New York: Vintage Books (2013).

Lin, Jean Z., et al. "Pharmacological activation of thyroid hormone receptors elicits a functional conversion of white to brown fat." *Cell Reports* (November 24, 2015): 1528–1537.

Louveau, Antoine. "Structural and functional features of central nervous system lymphatic vessels." *Nature* 523 (July 16, 2015): 337–341.

Maguire, E.A. "London taxi drivers and bus drivers: A structural MRI and neuropsychological analysis." *Hippocampus* 16, no. 12 (2006): 1091–1101.

Mann, Charles. *1491: New Revelations of the Americas Before Columbus*. New York: Vintage Books (October 10, 2006).

Moricheau-Beaupré, Pierre Jean, and John Clendinning (trans). "A treatise on the effects and properties of cold: With a sketch, historical and medical, of the Russian Campaign." Edinburgh (1826).

Morrison Shaun, Christopher J. Madden, and Domenico Tupone. "Central neural regulation of brown adipose tissue thermogenesis and energy expenditure." *Cell Metabolism* 19, no. 5 (May 6, 2014): 741–756.

Nestor, James. *Deep: Freediving, Renegade Science, and What the Ocean Tells Us about Ourselves*. New York: Houghton Mifflin Harcourt (2014).

Oelkrug, Rebecca, et al. "Brown fat in a protoendothermic mammal fuels eutherian evolution." *Nature Communications* 4 (2013): 2140.

Prince, V. H. "Double-blind, placebo-controlled evaluation of topical minoxidil in extensive alopecia areata." *Journal of the American Academy of Dermatology* 3 Pt 2 (Mar 16, 1987): 730–736.

Richalet, Jean-Paul, et al. "Physiological risk factors for severe high-altitude sickness." *American Journal of Respiratory and Critical Care Medicine* 185 (January 15, 2012).

Solomon, Christopher. "G.I. Joe and The House of Pain." *Outside* (April 13, 2011).

Steegmann, A. T. Jr., F. J. Cerny, and T. W. Holliday. "Neandertal cold adaptation: Physiological and energetic factors." *American Journal of Human Biology* 14 (2002): 566–583.

Tuttle, Alexander, et al. "Increasing placebo responses over time in US clinical trials of neuropathic pain." *PAIN, Journal of the International Association for the Study of Pain* 156, no. 12 (December, 2015): 2616–2626.

Umschweif, Gail, et al. "Neuroprotection after traumatic brain injury in heat-acclimated mice involves induced neurogenesis and activation of angiotensin receptor type 2 signaling." *Journal of Cerebral Blood Flow & Metabolism* 34, no. 8 (2014): 1381–1390.

van Marken Lichtenbelt, Wouter, et al. "Cold activated brown adipose tissue in healthy men." *New England Journal of Medicine* 360 (2009): 1500–1508.

van Marken Lichtenbelt, Wouter, et al. "Cold exposure—An approach to increasing energy expenditure in humans." *Trends in Endocrinology and Metabolism* 25, no 4 (April 2014).

Vosselman, Maarten J., et al. "Frequent extreme cold exposure and brown fat cold-induced thermogenesis: A study in monozygotic twin." *PLOS ONE* 9, no. 7 (July 2014).

Young, Andrew. "Homeostatic responses to prolonged cold exposure: Human cold acclimatization." *Handbook of Physiology, Environmental Physiology* (1996): 419–439.

ABOUT THE AUTHOR

SCOTT CARNEY IS an award-winning investigative journalist and anthropologist whose stories blend narrative nonfiction with ethnography. His reporting has taken him to some of the most dangerous and unlikely corners of the world. He is a senior fellow at the Schuster Institute for Investigative Journalism and a fellow at the Center for Environmental Journalism at the University of Colorado Boulder. He is the author of *The Red Market* and *A Death on Diamond Mountain*, and has been a contributing editor at *Wired*. Other works of his have appeared in *Mother Jones, Foreign Policy, Playboy, Details, Discover, Outside,* and *Fast Company,* among other publications. He lives in Denver with his wife, Laura, and their cat, Lambert. You can find more of his writings at scottcarney.com.

INDEX

Boldface page references indicate photos in the text. Underscored references indicate tables. An asterisk (*) indicates photos in the color inserts.

Breathing exercises *(cont.)*
  oxygen saturation from, 2–3
  pushups after, 3, 54, 56, 145
  sensations caused by, 1–2, 53, 55
  Wim Hof Method variation, 53
Breathing urge, 51–52, 89–90
Brown adipose tissue (BAT). *See* Fat,
    brown
Buijze, Geert, 116–21

C

Cancer, placebos versus drugs for, 131
Carney, Scott. *See also* Kilimanjaro
    expedition
  allergies suffered by, 49–50
  Buijze interviewed by, 116–21
  connection to the world by, 209,
    212, 215–16
  FAT and CHO oxidation rates,
    214, 215
  first lessons with Hof, 1–2, 8–12
  Hamilton visited by, 79–92
  Hof's house flooded by, 107–8, 115
  improvements in physiology of,
    212–15
  large calf muscles of, 45–46
  lessons learned by, 211–12
  MacKenzie visited by, 143–44
  metabolic baseline of, 44–45,
    46–47
  Snezka mountain climb by,* 17–19
  Spartan Race by, 65–66, 67–68,
    71–75
  Tough Guy run by,* 180–84, 185
  training program challenges of,
    108–9
  vision from holding breath, 88
  wayfaring sense loss by, 25–26
  workout routine of, 47, 48, 212
CO_2, 2n, 51–52
Cold exposure
  active conditioning, 62–63
  Army research on, 161–67, 169

in author's workout routine, 48
bodily reactions to, 12–13
brown fat activated by, 38–39,
    94–95
brown fat increased by, 35–36, 61,
    101
diabetes symptoms reversed by, 100
drug research on mice and, 96
in 15-minute daily workout, 64
guidelines for safe exposure, 61
Hamilton's work with, 84–85
handling the shock and pain,
    60–61
as Hof's teacher, 4–5
hypothermia due to, 160–61,
    163–64
MacKenzie on, 147
manual dexterity during,* 164–67
metabolism increased by, 61, 63,
    100
multiple benefits from, 98, 211
Nazi studies on death by, 12–13
November Project and, 150,
    152–54
pain when warming up after, 62
Parkinson's managed by, 127
during Ranger training, 160–61
research on Hof's method, 13–14
sensations caused by, 60–62
by soldiers in the field, 155–60
suppressing shivers, 61
swimming under ice, 7, 121
technique for practice with, 60–62
Tibetan monks' practice with, 13
van den Bergh's practice, 140,
    141–42
weight loss aided by, 40–41
Consciousness, 94–95, 97–98
Crohn's disease, 129–31

D

Daily workout, 15-minute, 64
Dean, Will, 174–75

Death, causes of
  acute mountain sickness, 110
  blacking out underwater, 89–90
  cold exposure, 12–13, 160–61
  drowning while swimming under
      ice, 121
  obstacle course races, 69
  tow surfing, 78, 90–91
Diabetes, 99, 100
Diamox, 110, 118, 188, 194
Dodge, Spencer, 160–61
Drugs
  altering fat metabolism, 99
  brown fat and, 94, 95–98, **96**
  Diamox, 110, 118, 188, 194
  for hair growth, 131–32
  limited usefulness of, 96–97, 101
  placebos versus, 131–32

E

El Nordico. *See* Gamester, Ed
Emmink, Hans, 129–31
Everest, Hof's attempt to climb, 8
Evolution. *See* Human evolution
Evolutionary mismatches, 33
Extreme Pool Training (XPT),*
    81–84, 147–49

F

Fainting or blacking out, 1, 55, 87,
    89–90
Fat, brown
  activated by cold, 38–39, 94–95
  activating on command, 63–64
  in babies versus adult humans,
      34–35
  drug research on, 94, 95–98, **96**
  Hof's method of activating, 14–15
  Hof's unusually large amount of, 14
  increased by cold exposure, 35–36,
      61, 101

in Neanderthals, 34, 35–36, 39
  white fat burned by, 14, 35
  in Wim Hof and his twin, 101–2
Fat, white
  in babies, 35
  burned by brown fat, 14, 35,
      95–96
  cold protection from, 38
15-minute daily workout, 64
Fire, evolution sparked by, 32–33
Freedivers, 28, 50, 87

G

Gamester, Ed,* 171, 172, 176, 177, 180
Gamester, Will, 176
Geronimo, 29–30
Graham, Brogan, 150
Guugu Yimithirr tribe, 24
Guy, King, 159

H

Hallucinations, 88–89
Hamilton, Laird
  author's visit with, 79–92
  breathing technique of, 86
  celebrities with, 80, 81, 82–83
  cold exposure work by, 84–85
  described, 81
  Extreme Pool Training of,* 81–84
  four-rule formula of, 92
  Malibu home of, 80–81
  on riding waves, 90–92
  spectacular career of, 79
  surfing the place of skulls, 77–78
  Wim Hof Method used by, 71,
      84–85
Hamis Ngonye, Salim, 207, 208
Headaches, altitude, 197
Heat acclimatization, 167–68
High-intensity interval training
    (HIIT), 144

Hippocampus, spatial thinking and, 27
Hitler, Adolph, 159
Hof, André (twin brother), 101–2
Hof, Enahm (son), 5, 109, 111, 112
Hof, Olaya (wife), 5–6
Hof, Wim,*
    bad business deals of, 111
    brown fat activated by, 14–15
    brown fat in his twin and, 101–2
    childhood of, 4
    claims made by, 2, 3
    cold as teacher of, 4–5
    complex character of, 3, 4
    development of technique by, 5
    Everest climb by, 8
    fame's effect on, 112, 113–14
    Guinness records set by, 6–7, 113
    "healthy, happy, strong" motto of, 114
    immune system control by,* 15–16, 102–3
    man saved from icy water by, 6
    new training center of, 111, 112
    raining in the house of, 107–8, 115
    scientific tests on, 13–14, 15–16
    son hired as manager by, 109, 111
    stunts by, 7–8, 9
    wife's suicide, 5–6
    woman's life saved by, 122
Holding the breath. See also Breathing exercises
    after breathing exercises, 1–2, 53, 56–57, 59
    extending the time of, 56–57
    in 15-minute daily workout, 64
    hallucinating due to, 88–89
    with lungs empty versus full, 54, 56
    recovery breath after, 57
    rolling muscle contractions with, 56–57, 59
    swimming with empty lungs, 87–88
    visualization meditation with, 59
Homo erectus, 32

Hot environment, 14–15, 63, 64
Human evolution
    dental health and, 33
    evolutionary mismatches, 33
    Homo erectus, 32
    Neanderthals, 31–32, 33–34, 35–36, 39
    technology's contribution to, 32–33
    triggering evolutionary programming, 41
Hyperventilation, 53. See also Breathing exercises
Hypobaric hypoxia, 118
Hypothermia, 160–61, 163–64

I

Immune system
    blood-brain barrier and, 104–6
    healing speeded by breathing exercises, 133, 134–36
    Hof's control over,* 15–16, 102–3
    Wim Hof Method's influence on, 103–4, 139–40
Indigenous peoples
    biological strategies for staying warm, 36–38
    loss of untouched populations, 37
    Pilgrims' contact with, 39–40
    psychic abilities in, 30
    wayfaring sense in, 23–25
Inuit, 36–37

K

Kalahari Bushmen, 37
Keneally, Scott
    about, 172–73
    Tough Guy documentary by, 173, 175–76, 184
    Tough Guy run by,* 181, 182, 183–84

Kilimanjaro expedition, 187–209
  author invited by Hof, 115
  author's request to join denied, 111
  boulder scrambling near the top, 208–9
  buddy system for, 119, 193
  Buijze on earlier climb, 116–21
  cell phone reception, 187, 189
  checking O_2 levels on, 193–94, 198
  Coca-Cola route chosen for, 188
  crew makeup for, 189–90
  dangers, 110, 117, 118, 121, 191, 192
  first casualty of, 199
  forecast for AMS in participants, 191
  Hof's mixed leadership on, 190–91, 193, 199, 200, 204–6, 209
  Hof's plans for, 110
  Hof's previous success rate, 189
  leg 1, to Mandara, 194–95
  leg 2, to overnight huts, 195–99
  leg 3, to Kibo and lunch, 199–204
  leg 4, to Gilman's Point,* 204–9
  rain and rain gear during, 195–96
  record-breaking goal for, 190–91
  "slowly" recommendation for, 188–89
  technique for headaches during, 197
  van den Bergh's advice, 140–41, 142
Kuze, Janis,* 9

L

Lake Louise Score (LLS), 119, <u>120</u>
Lange, Clive, 178
Lapps, 36
Lescelius, Andrew,* 2
Light, visions of, 1–2, 88–89, 145

M

MacKenzie, Brian
  assault bike training of, 145–46, 147
  author's visit with, 143–44
  controversy about, 144
  Extreme Pool Training of,* 147–49
  Wim Hof online course taken by, 144–45
Malik, Mahdi, 171
Mandarac, Bojan,* 150, 153–54
Manual dexterity, cold and,* 164–67
Master switch, 28–29, 50–51
McGinley, John C., 80, 82–83
Meditation, 57–59
Metabolism
  baseline for the author, 44–45, 46–47
  cold exposure increasing, 61, 63, 100
  of ducks, suppressed by cold water, 27
  of fat, drugs altering, 99
  passive increase in, 37, 38
Minoxidil (Rogaine), 131–32
Mr. Mouse. See Wilson, Billy
Muscle contractions, rolling
  brown fat activated by, 63–64
  heat generation by, 14–15, 64
  when holding the breath, 56–57, 59

N

Napoleon's invasion of Russia, 155–58
Navigational abilities, 23–24
Neanderthals, 31–32, 33–34, 35–36, 39
Nervous system
  blood-brain barrier, 104–6
  energy burned by the brain, 57
  the immune system and, 104–6
  parasympathetic, 55–56

Wilson, Billy (Mr. Mouse),* 173, 176, 177, 184
Wim Hof Method. *See also* Cold exposure; Wedge between autonomic and somatic nervous systems
for avoiding acute mountain sickness, 191, 192–93
breathing exercises, 1–2, 53, 55–57, 59
for combating altitude headaches, 197
confusing science for, 109–10
Crohn's disease and, 130–31
ease of learning, 109
15-minute daily workout, 64
Hamilton's practice of, 71, 84–85
healing speeded by, 133, 134–36
immune system influenced by, 103–4
MacKenzie's practice of, 144–45
obstacle course races for training in, 70–71
Parkinson's managed by, 2, 15, 127–29
popularity of, 111–12
pushups after breathing exercises, 3, 54, 56, 145
rolling muscle contractions, 56–57, 59, 63–64
van den Bergh's practice of, 139–40

X

XPT,* 81–84, 147–49